RIPPLES

RIPPLES

The Best of "Have You Ever Wondered?"

Geoffrey K. Watkins

Writers Club Press
San Jose New York Lincoln Shanghai

Ripples
The Best of "Have You Ever Wondered?"

Writers Club Press
an imprint of iUniverse, Inc.

For information address:
iUniverse, Inc.
5220 S. 16th St., Suite 200
Lincoln, NE 68512
www.iuniverse.com

ISBN: 0-595-22601-9

Printed in the United States of America

To Pat, because if it wasn't for you, it wouldn't be.
To Megan because you wrote me my first and only fan letter.
And to Erin, for putting it all on paper, again.

RIPPLES

Have you ever wondered about rain? Probably not since you were a child and I would guess not much even then. If you were anything like me, you were too busy enjoying it to waste much time thinking about it. Remember how much fun it was when it rained during recess or when you were outside playing and it took awhile for the grown-ups to realize that it had started? It was like stolen time. You'd put your head back and let the raindrops fall onto your tongue. But, you had to be fast, faster than any adult could call and order you inside where it was dry and boring. You were quick in taking advantage of those precious wet moments because you knew that once inside, all you could do was sit by a window, if you were lucky, and watch the rain drops softly pelt those puddles and create their tiny short lived ripples.

Do you remember? Some ripples were larger than others and those would quickly overtake the smaller ones, but they all seemed to merge effortlessly into each other, like the individual notes that make a fugue, only this was a fugue for the eye, not the ear. A frenzied conductor, you would jump in those puddles, creating in each one a great ripple that was soon overtaken by a myriad of smaller ones as the rain continued to fall, but once you were hurried inside and the door closed behind you, your silent symphony was over for the day. But there would be other days, other rains, until, probably about the time you left grade school, you stopped splashing in puddles. Still later you stopped watching the ripples altogether. But why?

We can lay some of the blame on the Greeks. Much of our notion of what schooling is all about comes from them. Even our word for formal education, academics, comes from the name of the small grove where Plato held his classes, Academe. The Greeks taught the inherent value of knowing the how and why of things and that was part of their legacy to us. So, school was surely a big part of the reason that you stopped splashing in puddles. Learning why and how something happens probably took a lot of the mystique out of it for you. Take rain as an example. I'd be willing to bet that anyone of us could recite now the basics of the water cycle; Water evaporates, rises as a gas, then condenses into clouds from which rain falls to start the cycle all over again. No magic about that surely. Just simple natural science. We all learned about that in grade school. The older we got, the more in depth the explanations became till finally, by the time we graduated from high school, we had learned all there was to know about rain. The problem was that we had also learned it was not dignified to splash in puddles. It was childish. Yet, the Greeks didn't teach us that, we added that sober notion all by ourselves.

You see, the Greeks had a rather more poetic way of looking at life than we do. They had a word, mousa, which gave us the word muse, to meditate, and Muse, referring to the goddesses of the arts. Distant though we are from the Greeks, we still refer to the Muses and generally we think of them somewhat like the Blue Fairy using her wand to grant Pinocchio his hearts desire, but the Greeks had a bit of a different mental image. To them it meant to throw back your head and allow creativity to fall into your mouth like rain. Less poetically, we just call it inspiration and think its something that happens to geniuses but not to us. We have to get by with perspiration.

So? And is there anything wrong with working hard and trying to know all the answers? Of course not. Ignorance is a sure path to childlike amazement, but it is still ignorance and therefore limiting. Knowing all the answers, however, is only the start. To gain understanding you have to open yourself to inspiration, to uncovering the true magic of life. And inspiration begins with wonderment. Once we begin to wonder, once we start to question, and do it with creative intelligence and knowledge, then we begin to see the interrelatedness of all creation, the ripples that make up our universe.

So, join me if you will. Put your head back and let the raindrops fall onto your tongue. Start where you like, stop when you want, but listen to nature's forgotten music. We are made of the same stuff as the dirt we walk on and the stars we aspire to. All of creation exists as ripples in that firmament, even us.

Have you ever wondered if there is other intelligent life in the universe? Outside on a clear night, looking up at the stars, it's hard not to be somewhat awed by the possibility. Here in the suburbs we can't see more than a fraction of the 3,000 some stars otherwise visible in our hemisphere, but that handful are generally enough to set minds wandering. But to where?

The nearest star to us is, of course, old Sol, our sun—some 93 million miles away. But beyond him, the next closest star to us is Proxima Centauri, approximately 4.3 light-years away. That doesn't sound like much until you consider that a light-year is literally the distance light travels in a year, at a speed of 186,282 miles per second. Put in perspective, that means that a beam of light could travel around the earth nearly seven and a half times in the second it takes you to snap your fingers. How far then did it travel from Proxima Centauri in the 4.3 light-years it took to reach us? About 25 million million miles. A jaunt around the park in a universe where the closest galaxy to our own Milky Way galaxy is 200,000 light-years distant.

But what about those neighbors of ours? Are they, or aren't they? The question is presently being probed by the SETI (Search for Extraterrestrial Intelligence) Institute and NASA. By scanning the radio spectrum of space, just as we might tune our car radios while traveling, they hope to catch an extraterrestrial signal from within our galaxy. Not an easy task, but the odds seem to be in our favor.

Frank Drake, the astronomer who founded SETI, has broken it down like this. With 400 billion stars in our galaxy, figure 10% (40 billion) have orbiting planets. But keep in mind, popular science notwithstanding, that astronomers only discovered planets outside of our solar system in the mid 1990's. Up till then, they were pure conjecture. So lets be conservative. Only take 10% of the total number of suns which might have planets and that means that of the planetary systems in our galaxy, there are 4 billion planets capable of producing life. Intelligent life? Let's be conservative. Say there is one chance in 15 million, lottery odds. That gives us 267 possible neighbors. And the kicker is, we're only talking about our own galaxy, one of 100 billion galaxies in the universe.

Looking up at that night sky, pondering the sheer immensity of it all, it would seem then that the question is not really do other life forms exist out there. In a house so large it is hard to believe we are truly home alone. The question is; Are we separated by distances so great that we will never be able to visit each other? If so, it would be truly tragic.

Have you ever wondered where computers came from? Like me, you're probably a bit computer phobic, or at least computer wary. When I was growing up they were the stuff of weird science and futuristics. Now they're everywhere: at our jobs, in our stores, in our homes, in our cars, and in our hands. I'm not talking about those lap-top computers we hear so much about, but just that hand held calculator you keep somewhere in the house to do those pesky little math problems. It's also a computer. And a powerful one at that. But how did it get there?

It all started in 1650 when a young man named Blaise Pascal invented the first computing machine. It manually added and subtracted numbers by using toothed gears, the same principle that operates your car's odometer. About twenty years later Wilhelm Von Liebnitz developed a machine that could also divide and multiply. With that, computers were on their way.

Fifty years later, these two ideas were wedded to each other in the fertile mind of Charles Babbage, an eccentric Englishman. Assisted by Ada Lovelace, daughter of Lord Byron and a gifted mathematician, Babbage designed two computer-like devices, but failed to actually produce either due to the inadequate technology of his time.

In 1890 Herman Hollerith put it all together with his invention of a working electromechanical device that could take directions from punched cards. His machine's worth was proven by its completion of the 1890 census in three years, not the projected ten. In 1924, Hollerith's company became IBM.

We begin to chart the generations of modern computers with the first completely electrical computing device, the 1941 Atanasoff-Berry Computer (ABC). The best-known offspring of this first generation was, however, the WW II era ENIAC. Developed for the military, its actual usefulness was limited by its size and the power demands of its hundreds of vacuum tubes. Beginning in 1959, transistors replaced the vacuum tubes and the second generation was born only to be left behind five years later by the introduction of integrated circuits. This third generation has seen the development of even smaller, faster and more reliable computers like your calculator. The fourth generation? Don't blink.

Have you ever wondered why your ears sometimes pop? When I was growing up, my mother always carried gum in her purse if we were going to the mountains. On the way home my hearing was less acute than it had been and my ears would feel as if I had a bad head cold. Out came the gum and a few minutes later, pop! But why? The answer was right above my head.

Outside on a clear, cloudless day, you can look up and see nothing but sky. The reason you see nothing is because all that is above you is odorless, colorless gas. We generally refer to the whole of it simply as air, but it is actually a rather complex mixture of Nitrogen, Oxygen, Argon and trace elements of a number of other gases. Without that recipe in the right proportions, life on earth would soon cease, but consider how fragile that mixture, our atmosphere, is. The higher one rises the thinner the atmosphere becomes, until at around 250 miles above the surface of the earth the molecules that make up air are so far apart as to be virtually immeasurable. Now, 250 miles of air seems like a great deal, but, considering the size of the earth, it really isn't very much. It is approximately the thickness of a coat of varnish on a world globe. And precious little of that 250 miles is breathable air. So little, in fact, that all life exists within the bottom three miles of our atmosphere.

Above us then lies a veritable ocean of air of which we tread the very bottom. By contrast, the deepest known point of any terrestrial ocean is Challenger Deep, nearly seven miles below sea level. At that depth, with the weight of seven miles of water pressing down on him, a human being would be immediately crushed. Yet, with 250 miles of air above us, we live quite comfortably. Because air has no weight? Think again. In fact, air does definitely have weight. A handful of air weighs about as much as an aspirin. At this very moment, the atmosphere above your head is pressing down on you at the rate of 14 pounds for every square inch of your body, about 15 tons all told. So why aren't we crushed? Because your body makes sure that the pressure inside is equal to the pressure outside.

You see, behind each eardrum you have a tiny valve called the Eustachian tube. It regulates the air pressure inside your body with the air pressure outside by opening and closing. When you ascend, the higher pressure inside pops the tube open and equalizes the pressure inside and out. Descend and the opposite occurs—the greater pressure outside clamps the tube closed and the only way to pop it open is by taking in air by yawning or swallowing. Hence the reason gum helps. Don't tell Dad, but Mom always did know best.

Have you ever wondered where birds come from? For as long as recorded history, and likely before that, people have envied birds their freedom of flight. The ancient Greeks subtly warned against that envy with their cautionary tale of Icarus and his wax and feather wings. In the 16th century, Leonardo Da Vinci sketched a design for wings that could be donned by a human. Finally, in our century, the Wright brothers achieved manned flight, appropriately enough at Kitty Hawk, North Carolina. But, for all that, we still lack the subtle grace of a bird. Then again, they've been at it longer. In fact, at least 150 million years longer.

In 1861, workers in a limestone quarry near Solnhofen, West Germany, discovered the fossil remains of a small creature that would send tremors throughout the scientific world. It was only 20 years before that Richard Owen, a young British scientist, had first proposed the existence of a new class of ancient land reptiles. He named this new class Dinosauria, "terrible lizards." But this new find was different. It appeared to be neither terrible, nor a lizard. Rather, it looked like a bird, albeit a bird with teeth. Of the five specimens found in Solnhofen, four show the distinct impression of feathers. And, for the modern paleontologist, there is the rub.

Take away his feathers and this first bird, Archaeopteryx, is remarkably similar to a contemporary of his; a small dinosaur called a Compsognathus, from the family of dinosaurs called coelurosaurs. Were they related? It depends on whom you listen to. Paleontologists agree that birds descended from thecodonts, the ancestors of dinosaurs, but the traditional view holds that this family split into two distinct branches.

One branch led to dinosaurs, the other to birds. Similarities between the two are seen as the result of parallel evolution, what scientists call convergence. In 1969, however, Professor John Ostrom of Yale University called its application in this case erroneous. Looking at the myriad similarities between Archaeopteryx and Compsognathus, Dr. Ostrom postulated that both animals are in fact coelurosaurs. They were both dinosaurs.

Are dinosaurs really extinct? Perhaps not. Maybe those small creatures in your yard are the direct descendents of those "terrible lizards" of our best dreams and our worst nightmares. Maybe, as Dr. Ostrom has said "Dinosaurs didn't become extinct. They simply flew away." And who knows? Maybe, in the far distant future, an alien archaeologist might well write the same about humans. Maybe.

Have you ever wondered about New Year's resolutions? For most of us, New Year's Day is a day of reflection following a night of revelry, of "ringing in" the upcoming year, but where did it all start? It depends on where you look.

Go back five thousand years to ancient Egypt and the new year began in early summer, with the annual flooding of the Nile and the subsequent fertilization of the surrounding delta. From time immemorial the Jewish people have celebrated the new year in the early fall, around the time of the autumnal equinox. It is in ancient Rome, however, that we see the first glimmerings of our New Year customs.

Prior to 700 B.C., the Roman calendar had only ten months. To make the calendar more accurate, Numa Pompilius added two new months, January and February, to the end of the calendar year. He named January after the Roman god Janus—the god of gates and doors and beginnings and endings. Appropriately enough, Janus is usually depicted as having two faces, one looking forward, the other backward and was thus like the Roman people who, on the cusp of the new year, looked to what had been in the year passing and wondered what the coming year might bring.

To sweeten the pot, as it were, Romans would visit on the new year's first day and give each other gifts of flowers, fruits, or other tasty or tasteful items. Unfortunately, however, as time passed, the calendar of Numa Pompilius, though initially more accurate, became less so and the actual—that is solar—new year became harder to fix. The problem was effectively solved in 46 B.C. by Julius Caesar.

Among other corrections, the emperor made January the first month, added a day to it, and placed the first day of the new year on January 14. The Julian calendar remained the standard until the introduction in 1582 of the even more accurate Gregorian calendar, the calendar we follow today. It was this calendar of Pope Gregory the XIII which forever fixed January 1 as the initial day of the new year.

But what of our custom of making resolutions? Well, is it really so different from the ancient Roman's practice of looking at the past and towards the future while bestowing gifts? Isn't a resolution to make the new year better really a gift to ourselves and others? Perhaps then the best new year's gift we can give is the resolution to follow the simple precept of John Wesley: "Do all the good you can, by all the means you can, in all the ways you can, to all the people you can, as long as you ever can." That would indeed make for a joyous 1993.

Have you ever wondered how the notion of saving daylight came about? Well, contrary to what you might think, it wasn't an American invention, though the United States has made the most of it since it got started back during World War I.

At that time, England, France and Germany were the first countries to see the advantages in having an extra hour of work time each day; not to mention the energy savings. America caught onto the idea near the war's end in 1918, but Congress rethought the whole thing and finally repealed the law in 1919. Not that many people really cared. The twenties were a time of growth and consumption; economy was not the order of the day. The thirties were not so flush, but with much of the country out of work and production down, who needed an extra hour of light? No one. Until World War II.

From 1941 to 1945, the United States was on a program of year-round fast time aptly named "War Time", but a year later, what time it was truly depended on where you were at the moment. While some states went back to Standard Time after the war, others liked the idea of Daylight Savings Time, while still others allowed individual counties or even towns to decide whether or not to advance their clocks!

To totally confuse the issue, there was not a set beginning or ending of Daylight Savings Time. That state of confusion lasted for twenty years. It was not until 1966 that the Uniform Time Act set the official dates for Daylight Savings Time and required all states to adopt it unless their legislatures voted against it.

Finally, in 1986, the beginning of Daylight Savings Time was changed from the last Sunday to the first Sunday in April. What's next? Only time will tell.

Have you ever wondered about color? As the song goes, "birds do it and bees do it," and yes, even fleas do it, educated or not. But, even an educated dog cannot do it. Do what? See color. Matter of fact, lots of insects, bees for instance, see colors we can't. Which is something we should all be thankful for because their life, and by extension ours, depends on that ability.

To our eyes a flower is mostly a mass of colored petals. To a bee, the color that matters is located in the center of that flower, precisely at the point where the nectar collects. That color, ultraviolet, is invisible to us, but it's a homing beacon to a bee. Once the bee identifies the ultraviolet spot he lands on it, collects the nectar and the pollen that is also there and then moves on to another flower, another plant, and thus pollinates a world of fields and orchards. Without the bee, life as we know it would not be. But without that tiny spot of color, life as a bee knows it would be equally impossible. But what exactly is color? Well, it is a component of light.

In 1666, Sir Isaac Newton projected a beam of ordinary white light through a prism and studied the now familiar rainbow effect we see in gift-shop crystals. With that rainbow, called a spectrum, Newton proved that ordinary light is actually made up of a myriad of different colors. He then postulated that light is made up of tiny particles, corpuscles, which travel in unerringly straight lines. He was half wrong.

Christian Huygens, a Dutch contemporary of Newton, noted that the water in a pond is deflected around a rock, not stopped. The same, he saw, is true of shadows. Their edges are not as distinct as their centers. This led to the correct conclusion that light behaves the same as waves in water, they bend. And thus the wave-theory of light was born.

Fine and good, you say, but what about color? Well, each color of the spectrum is a distinct electromagnetic wave. Think of it in terms of a radio, which also relies on electromagnetic waves. To hear a station, you must tune to the particular wave on which that station is broadcasting. To see a color, your eyes must tune in to the correct color wave. They do so picking out the separated light waves reflected off the object we are looking at. The shorter color waves, like the one our bee can see, but we cannot, are called ultraviolet. The waves, which are longer than we can normally see, are called infrared and are most familiar to us from medical infrared photography. Of the total range, however, humans can only see about 3 per cent of all light waves, or colors. Not much perhaps, but how dull life would be without that precious 3 per cent. It gives a whole new perspective to the old saying about a dog's life.

Have you ever wondered about gold? Sounds like a trick question, doesn't it? We live in California—the Golden State. Our national monetary system was built on the gold standard. Most of us own at least one piece of gold jewelry. But, how much do you really know about gold? Probably not as much as the Quimbaya, the Muisca, the Sinu, and other ancient South American Indian tribes did. They had been mining and working gold long before the Spanish conquistadors arrived in 1519. The Spanish came for spices, but they left with gold; tons upon tons of gold. Hard to imagine? Try this on for size.

Approximately 80,000 tons of gold has been mined in the past 6,000 years. That equates to a cube with six sides of 53 feet each or a football field covered by a slab two and a half feet thick. So where is all that gold now? Governments and central banks hold about 60% of it. Our government, for example, has a reserve of about 9,000 tons. The rest is owned by individuals in forms as mundane as gold bars, as precious as the wedding ring on your finger, or as unique as the solid gold inner coffin of King Tutankhamun; it is the largest surviving gold artifact from antiquity, weighing in at 244 pounds. But why do we place such worth on gold that we marry each other with bands of gold and bury Kings in coffins made of it? Because of its golden qualities.

Gold does not tarnish and is chemically more stable than copper, aluminum, or stainless steel. It is dense—one cubic foot weighs about 1200 pounds—and very malleable. So malleable, in act, that one ounce can be stretched into a one-half inch wide ribbon one and one-half miles long. That ribbon would be so thin that light could pass through it, but it would still reflect infrared rays, making it a near perfect window coating/insulator material. You have probably seen the famous picture of Neil Armstrong on the moon in which the Lunar Lander is reflected in the golden visor of his helmet. It is also a superb conductor of electricity. About 10—15% of current gold production goes into TV, computer, and telephone circuitry. Where else? Well, a large portion goes into medical and dental use, but by far the largest consumer of gold today—about three fourths—is the jewelry market. And the greatest use in the United States is class rings. Nearly 3 million of them each year. About 15 tons of gold.

Heard enough? Well, maybe one last tidbit to top you off. In India's traditional Ayurvedic medicine, gold is among the foremost of pure and auspicious objects. To partake of that purity, pills of gold are often prescribed. So, maybe that heart of gold you always wanted isn't so far away after all.

Have you ever wondered how writing developed? Living, as we do, in the midst of what is already being called the "information age" that might seem an odd question. Hasn't writing always been around? Well, not as we know it. For us, writing means using a series of abstract visual symbols-letters—to put our words and thoughts into a more or less permanent form. It is a code, which we can be taught to both decipher and transcribe. Every literate person in the world knows at least one of these codes, be it English, Italian, Hindustani, or any other writable language. Some, like Korean, are exemplary for the ease with which they can be learned. Others, like the 5,000 letter of the Chinese alphabet, are beautiful, but daunting. But the initial idea behind these expressive codes apparently sprang from the mundane need to keep track of one's belongings.

In the dim past, human beings likely had learned to keep a simple one-to-one tally using stones, shells or other convenient objects, but there was no way to tell whether a stone represented a sheep or a sheaf of wheat. Denise Schmandt-Besserat of the University of Texas thinks that the thousands of clay tokens, which have been collected from excavations in the Middle East, are the ancient answer to this problem. These tiny objects some 8 to 10,000 years old and approximately the size of your thumbnail, were made in fifteen distinct types. Professor Schmandt-Besserat has theorized that these abstract clay tokens represented real objects and were made for business accounting purposes. Used as bills of lading, these tokens would be sent enclosed in clay balls called "bullae". The balls were cumbersome, however, and had to be broken open for the tokens to be tallied. Until, that is, some accountant reasoned that it would be much simpler to press the tokens into the wet clay ball before it hardened. And with impressions of the ball itself, who needed tokens? Or bullae? A flat clay tablet was easier to work with. You didn't need tokens, just a sharp stick. The tablets were also much easier to transport and store.

By this time, people had also figured out that a one-to-one relationship—one mark, one object—was not very efficient, and so the concept of quantity versus object was hatched. They could now make a single distinct mark for any amount—a number—followed by an equally distinct mark for any object—a word: 5 sheep; 3 sheaves of wheat. Writing had been born. So you see, without the tax man's ancestors, we wouldn't have writing. Though we might have had a free lunch…

Have you ever wondered about robots? The desire to create mechanical life has been lurking around the fringes of human consciousness since the Greeks. They had simple gear driven toys that stiffly approximated lifelike movement, but it was not until the 18th century that Europe was beguiled by automatons and the modern ideal of non-organic life took root. Generally taking the form and dress of exotic Middle Eastern adults or children with the sweet look of innocence on their wax faces, these mechanical figures moved by an intricate and ingenious series of gears and pulleys. They were, however, likewise limited to those contrivances to a certain set series of movements, which could only mimic life, not participate in it.

Another plateau was reached in 1920 when the play R.U.R. was produced and the world began to consider the ethical questions behind the possibility of creating mechanical life. Written by Czech dramatist Karel Capek, R.U.R. referred to Rossums Universal Robots, the protagonists of his play. Indicative of their planned position in society, the word robot came from the Czech term, "robotit"—to drudge.

Finally, at the beginning of the atomic age, scientist and author, Isaac Asimov set forth the specific rules governing robot and human behavior and interaction. In the 27 years since their named introduction, human perception of robots had progressed to the point that they were no longer seen as mere drudges. They had attained those inalienable rights, which at our noblest, we extend to all intelligent life forms. But that life is, of course, a fiction. Real robots are more like automatons than CP30. They perform their functions as programmed, with no variance, weariness, joy, or interest.

Today, however, with computers has come the real possibility of creating an approximation of human intelligence, generally called Artificial Intelligence, or AI. Capek's and Asimov's robots have this capacity; they can think and so have life, but are they like us?

To stay alive, we must breathe. Fortunately, we do, but we don't consciously think about breathing, or digesting food or circulating blood, or any of the myriad other functions we need to do every second of the day. These functions are called autonomic and are like the repeated actions of real robots, ceaseless, programmed, and unvarying. Since these functions are automatic, we are freed to think and thus grow. Yet, how many of us fully utilize that gift? Perhaps then the question is not are robots like us, but, are we too often like robots?

Have you ever wondered about numbers? It is a safe bet that most of us are at least a little bit number phobic, but think of how difficult life would be without a symbolic—that is written—numeration system. Without number symbols, how could you ever figure out daily transactions like buying lunch, going shopping, or writing a check? A one-to-one correspondence? Three notches equal three dollars? Can you imagine what the national debt would look like if it could only be represented by individual marks? Life is far too complicated for that simple of a system. Truth is, it's been too complicated for that for a long time. So, where did it start?

Naturally enough, counting began finger by finger, with five objects equaling one "hand". Today we follow this ancient rule by making tallies of four straight marks crossed by one diagonal line to group by fives. The next step was to use both hands and thus group objects by ten. By such a system, the Egyptians were able by the year 3400 B.C. to signify numbers up to and exceeding one million. Their symbols were simple line pictures which represented the particular corresponding power of ten. For instance, to write the number one million, ten to the sixth power, or 10 X 10 X 10 X 10 X 10 X 10, our Egyptian accountant would draw a stick figure of a kneeling man with upraised arms, an astonished man. As you can imagine, unrounded numbers, like 2,437, made for some very complicated drawings. In contrast, the Roman system was far more logical, but it was by no means more practical.

The Romans relied on seven basic symbols which represented the numbers 1(I), 5(V), 10(X), 50(L), 100 (C), 500(D), and 1000(M).

With those seven simple symbols, the Romans were pretty well set up; unless they wanted to write large numbers. In one notable example a monument was erected to commemorate a victory over the Cartheginians. They showed the number vanquished—2,300,000—by repeating their symbol for 100,000 twenty-three times! What we can write with seven symbols, took the Romans 23. Why? Take a look. They had no symbol for the absence of a power of ten. Like the Egyptians, they had no zero. But, luckily for us, others did.

By the time of the fall of Rome in 455, the Mayas of the as yet undiscovered "new world" had developed the concept of zero. Across the Atlantic Ocean, in India, the Hindus had invented the number system which we use today, the familiar digits of 1 through 9, plus 0. From them the famed scholar merchants of the east, the Arabs, would soon bring more than silks and spices to Europe. They would bring the ease of enlightenment.

Have you ever wondered about time? It can rule our lives. It is inescapable. And depending on what you are doing, it can be a boon or a bane. But, as with most things in life, it is often what you make of it that matters, not what it really is.

At its most basic, time is one of the two primary dimensions in which we live our lives, the other being space. Take it a step further, however, as did Einstein, and the two become one inseparable dimension where objects, us included, exist in terms of length, width, and heighth as well as past, present and future. But for most of us, time is more intimate and immediate; we are generally running to meet the myriad connected moments which make up our lives. Yet who determines if we are being punctual or not? The government does.

Earliest man likely marked time by the passing of days or by the distance from major events, such as "since the time of the flood." Later came the Egyptian and then Babylonian calendars, useful for measuring days, months and years, but not for the smaller units of hour, minute, and second which we rely on today. But for the great majority of people in the world, the visible passage of sun time, the "apparent solar day", was enough. One rose with the sun, rested at midday, and stopped when the sun set. Only with the rise in technology brought about by the 19th century industrial revolution did the common man begin to need more precision in his life. In America, 1883 marked the beginning of our now chronic awareness of time. It started, however, twenty-one years earlier with a vision.

In 1862, Theodore Judah saw need and opportunity looming on the western horizon. He rightly reasoned that people would willingly pay to ride in relative comfort and safety to the new frontier. Trains were the answer. The problem was time.

Before 1883, trains relied on local "sun" time for their schedules. Since the sun first appears in the east, however, this means that when it is noon "sun" time in Washington, D.C., it is 9 a.m. "sun" time in Los Angeles. Conceivably, one could thus both leave and arrive at the same time! Under this system, havoc was the rule. In 1883, therefore, the United States government established the four standard time zones which simultaneously divided yet unified the contiguous United States and territories. Each day, each hour, a signal was broadcast from Washington, D.C., ensuring that going west, each zone would be exactly one hour behind the last and accurate to local sun time. So, thank you, Mr. Judah! I guess...

Have you ever wondered about sound? Remember that old riddle about the tree falling in the forest? If no one is there to hear the tree fall, would it make a sound? Well, yes and no. The answer depends on how you define sound.

Simply put, sound is something you hear, but how you hear is a story in itself. Hearing is actually a physical reaction to a series of vibrations that have passed from outside to inside your head. Specifically, the vibrations are first gathered by your outer ear and channeled through your ear canal to your eardrum. There, like talking drums, the sound/vibration is transferred to the three smallest bones in your body. Taking up the message, these three bones in turn stimulate fluid in the cochlea—a tiny structure in your skull that looks for all the world like a soft snail—causing the 24,000 fibers packed in its one and one-quarter inch length to stimulate the specialized nerve cells which will then carry the vibration to the temporal lobe, the center of hearing in the brain. Stop the vibration anywhere along the way and hearing is prevented. But what about sound?

As we said, sound is vibration. Hit a bell and you can feel the wall of the bell moving rapidly back and forth. That tiny movement causes the surrounding air to vibrate at a certain frequency, or pitch. As the air is pushed outward by the expanding object, the surrounding air is compressed. As the object then rebounds, air rushes back in to fill the void. This series of compressions and expansions—scientists call them condensations and rarefactions—travel outwards, as waves of sound, until they simply fade away.

So what about our falling tree? Does it produce a sound? Well, it still depends on a couple of factors. In this case, density and elasticity. Remember the old cowboy movies where the Indians put their ear to the ground to hear better? They used a simple principle; sound waves travel through different mediums, substances like air or steel, at different rates. The denser the medium, the slower the speed. The more elastic, the faster. Thus, though steel, for instance, is 6,000 times denser than air, it is nearly two million times more elastic. A sound wave travels about 17,000 feet a second in steel versus 1,200 feet a second in air. The sound is also strengthened. Remember that a sound wave loses intensity as it moves along. If our organ of hearing, the ear, is too far away from the source of the sound and the density and elasticity of the medium are not sufficient to carry it to us, we do not hear the sounds. So, if no one is within range of hearing when the tree falls, does it make a sound? Well, yes and no.

Have you ever wondered just how much a million is? As with most things in life, a "million" is pretty relative. It all depends on what you're talking about.

Remember Mr. Potter from "It's a Wonderful Life"? He was a millionaire. He had a million dollars socked away and seemed to live pretty well, even if he was a bit mean spirited about it all. Nowadays we tend to yawn when the Lotto jackpot is under three million. How about the original "Beverly Hillbillies"? Jed was a millionaire. These days, in the current movie remake, he's a billionaire. Well, you're thinking, back when Mr. Potter and Jed were current fare, money was worth a lot more. Granted. Life just doesn't seem to be what it used to be. A million dollar bills just don't excite the human imagination like they used to, but consider the following: If Mr. Potter got bored one rainy day and decided to stack his one million dollar bills, the pile would be 667 feet high. If a million people decided to go see "The Beverly Hillbillies", the line would be nearly 200 miles long. Put all those people back in their cars after the movie and send them home on a three lane highway and the traffic would be backed up from New York to Miami. Boggling, isn't it? When you get right down to it, a million is not nearly as trifling a sum as we think it is, but as I said earlier, it is all relative.

Consider, if you will, the paper you are holding right now. It, like the hand holding it and every object around you, is composed of matter. These aggregate bits are in turn made up of atoms, the smallest particles of matter with distinct chemical characteristics.

Picture a sand castle. To our eyes it is solid, yet it is made out of millions of distinct, yet cohesive grains of sand. As a rough analogy, the same is true of matter and atoms, but on a different scale. Take just one grain of sand from your imaginary sand castle and lay it in front of you. Now, take all the atoms from that one grain of sand and make each of them the same size as the grain of sand you started with. Ready for this job? With those enlarged atoms, you can now make a sand castle one mile high, one mile wide and one mile long. Hard as that might be to imagine, the actual size of atoms is even more difficult to comprehend. The smallest atoms, those of hydrogen, are in fact so small that one million of them lined up side by side would not equal the thickness of one of Mr. Potter's dollar bills.

Maybe life isn't what it used to be. Maybe a million isn't what it used to be. Or, maybe neither of them were ever what we thought they were to begin with. As I said, it's all relative.

Have you ever wondered what the most important life form in the world is? Most of us haven't because of the value we place on our own continued existence. The answer is self-evident; we are. Perhaps a more thoughtful answer, however, is that since all life forms are interdependent on each other, there can be no "most important." But, mightn't there be degrees? Would you say that an earthworm is as important as a human? Would you say that algae is as important as a whale? Look a bit closer at the matter. Just how do we perceive importance?

In many eastern cultures age is akin to importance. To a Westerner, size might be a factor. Blue whales satisfy both criteria. They live up to 50 years, nearly as long as humans and at 100 feet in length, they are the largest animals to ever grace the earth. Are they important? Of course. All life is important. But are they personally vital to you? Does your life or your families' immediate or long-term well being depend on blue whales? Likely not. Does its' longevity or size therefore make it a contender for "most important"? Of course not. Surely it is not age or size, but actions that count more in terms of importance. So where do we look now? Start below your feet.

Even though the common earthworm can attain a length of up to eleven feet and live upwards of twelve years, most people find them a lot less impressive than a blue whale. Without their digging, however, the soil on which land animals ultimately depend for sustenance would be as sterile as moon rocks. No worms, no fertile soil, no plants, no animals, no life. On the land then, size and age wouldn't seem to count for much in terms of relative importance. But that's the land. What about the oceans?

Well, that takes us back again to our blue whale. He is, you see, rather remarkable not only for his size, but also for his eating habits. Each day he consumes tons of microscopic one-celled bits of animal and plant life collectively called plankton. These tiny bits of life form the lowest rung of the oceanic food chain; the rung on which all sea life depends. No plankton, no oceanic life. But they don't stop there. Plant plankton, what most of us think of as algae, has another wonderful function. It produces about 80% of the world's oxygen supply. Without these tiny unheralded organisms, our world would not be able to sustain oxygen breathing life. In short, we would not be here. The most important life form on earth? Human beings, of course. But on New Year's Eve, I celebrated those organisms that make it all possible. Happy New Year to us all.

Have you ever wondered about ice? Not the most scintillating subject, right? I mean, how much is there to really consider? Well, a lot really. You just have to know where to look. Let's start close to home.

Take a peek in your freezer and you'll find ice. Not much of a surprise there. It's cold, probably around 5 degrees, 27 degrees below freezing, so you'd expect to find ice. But consider what a minor miracle that really is. Even on the coldest day of the year I'd be willing to bet your ice cube tray would have nothing in it but water if you left it on your kitchen counter for an hour or so. Yet, put it back in the freezer, ice again! Frozen water. But frozen fresh water, not salt water. You see, salt water doesn't freeze, it just turns to slush. So what about icebergs? They're frozen water, right? Well, no. Let's go exploring a bit farther from home. But, let's start at the beginning, let's go back to the father of all things, Aristotle.

The Greeks were sticklers for balance. To keep the universe on an even keel, everything had to have its polar opposite, literally. Aristotle taught that since there were lands to the far North, there must also be lands to the far South. Balance. And since lands to the North were grouped under the constellation of Arktos, the bear, so the lands to the South must be under a similar, but opposite constellation, Antarktos. The name has come down to us as Antarctica. And if it's ice you want, this is the place.

Remember the seven continents? Quick now, which is the largest? Asia. Which is the smallest? Australia. Which is the highest? Antarctica. Surprised? You're not alone. Most people would say South America because of the Andes, or Asia because of the Himalayas, but they'd be wrong. You see, Antarctica is primarily ice, fresh water ice, some seven million cubic miles of it. In fact, this fifth largest continent contains almost 70 percent of the Earth's fresh water. So what, you ask? Well, that ice covering has an average thickness of around 7,100 feet, nearly a mile and a half in height. That gives Antarctica an average height above sea level of around 7,500 feet, higher than any other continent by far. But, that's the continent itself, you say, what about icebergs? Well, it's kind of like not being able to see the forest for the trees.

Imagine if you will that you are standing deep in the interior of Antarctica. You needn't worry about being eaten by polar bears, they don't live here. Matter of fact, Antarctica has no resident large animals and only one small one, a species of fly. The real danger here is the cold. The lowest temperature ever recorded on Earth was here, 128.6 degrees below zero. Far colder than it need be for snow, but Antarctica lives up to its name, it is a land of extremes. It does snow there and that snow is forever being pressed down by yet more snow and once pressed it condenses into ice. From the enormous weight from above, this ice forms frozen rivers that are pushed along towards the outer edges of the continent at almost observable speed, up to half a mile a year, till they reach the edge of the continent and having nowhere else to go, they float out on the open waters of the ocean. These floating edges of ice, properly called ice shelves, are not insignificant in size. The largest, named the Ross Ice Shelf after the famous 19th century explorer, is as large as France and nearly one half a mile thick. Considering the tremendous strain these huge sheets of ice are under, it should come as no surprise that as they get pushed farther and farther out to sea, chunks begin to break off of the leading edges in a process called calving. Once in the open water, they are on their own as icebergs, some of which can be as large as 5,000 square miles, the size of Connecticut. But they do not remain so large as that forever. Evaporation takes its' toll as they drift away from their parent continent and the once frozen water rises as gas into the atmosphere. There it condenses and forms clouds which produce snow for Antarctica, but rain for the more temperate climes where we live. And when that rain falls it collects in lakes and reservoirs and aquifers where it is piped into our homes. And we take that water and pour it into our ice cube tray and place it in our freezer, knowing that in a short time we will have fresh ice to cool our drinks. Quite a journey, don't you think?

Have you ever wondered about muscles? Maybe a better question would be who hasn't? As a country, we love to exercise our freedom to show how well we fit the perception of what we all should look like; lean and mean. Clint Eastwood embodies that look for us. He's tough. He's strong. He's in control. He's worth the price of a movie ticket. But, how much of that difference between him and us is real and how much is our perception?

Well, to begin with, we all have the same number of muscles in our bodies, over 600, with the greatest number of muscles being in the face. That concentration allows one of the marked differences between humans and the lower animals, our ability to show a wide range of emotion and communicate by facial gestures alone. Just think about a Clint Eastwood western and you're likely to remember his face in close-up and what it told you was going to happen next. He's tough, but he's not inscrutable. You're perceptive enough to read him.

How about strength? Well, it's another dead heat. You see, your skeletal muscles, the ones attached to your bones, can get bigger by exercise and thereby make you physically stronger, but you'll never be Superman. The problem is, your muscles must work in pairs to move your skeleton, each muscle pulling in a different direction. If all the muscles in your body could be made to work in the same direction at the same time, you could lift about 14 tons, roughly seven automobiles. Clint could do about the same. So, it's your perception again.

Let's get back to that close-up. Clint's face muscles are set. Your blood pressure rises a bit and your stomach muscles tighten as you watch him. Actually, Clint's "smooth muscles"; his stomach, intestines, and blood vessels, are doing the same thing as yours. Without us think-ing about it, our smooth muscles all work to supply our skeletal mus-cles with the chemical energy they need to respond instantly to commands from our brain. Commands that will travel some 300 feet per second. That is real control, and we all have it. Once again, it's our perception that makes us think Clint is different.

One last muscle, one last chance. Your cardiac muscle, the heart. In one hour it expends enough energy to lift a five-ton weight more than a foot off the ground. Just like Clint's. No difference there. And his beats as often as yours, around seventy times a minute, day in and day out for a lifetime. So, in the end, there are no real differences between us. It's only exception that makes Clint a heart-throb and each of us some-body's sweet heart on Valentine's Day, but, thank God for that percep-tion!

Have you ever wondered about leprechauns? I would guess not recently. Age has a way of drumming the magic out of our lives, but think back to when you were smaller and the world was larger. Didn't you dream more then about life's possibilities?

Leprechauns? Of course there are leprechauns. They're responsible little people who work every day making shoes for their still smaller cousins, the shees, or fairies. They also have other less desirable traits: they're cranky loners who hoard gold and will do anything to keep it. Akin to dwarfs, leprechauns are earthy creatures, not much given to philosophizing and are thus doubly betrayed by the airy rainbow whose beauty they do not appreciate and whose end falls on their hiding places. Find the rainbow's end and you'll find a pot of gold. But, be careful. It's fairy gold, you see. And fairy gold is hard to grasp and harder to keep.

At least that is what I was told as a boy. What I learned was that the rainbow's end is a very elusive destination. It seemed that I could never walk fast enough or far enough to please the fickle rainbow before it got tired of waiting for me and left for home. But where was home? Though I didn't know it then, it was all around me. You see, sunlight is actually composed of seven blended colors: violet; indigo; blue; green; yellow; orange; and red. When sunlight passes into raindrops the water bends, or refracts each individual ray and separates each color in that ray. These colors are then bounced back or reflected out of the raindrop where they are further refracted and dispersed in such a way that the colors we finally see are determined by the angle at which the sunbeams initially struck the water droplets. Red, for example, occurs at around 42 degrees above the line of the sun's rays. All the other colors we see occur at lesser angles. Think about it. The higher the sun goes, the lower the bow, till finally, when the sun rises above 40 degrees, its rays drop so steeply that there can be no rainbow. Likewise, without the correct angle between you and the sun's rays, you cannot see the rainbow. Walk towards it and you lose it. Keep your perspective and you keep the magic.

So, while I don't hear much playground wisdom these days, I nonetheless believe in leprechauns. I see them everyday; people who have abandoned their pot of gold at the end of the rainbow for a light at the end of the tunnel. As for me, I have no wish to reach the end of my life only to find that the gold I worked so hard for was fairy gold, illusory, and that while I was grasping for it, I let life's real magic slip through my fingers. Anyone up for a rainbow break?

Have you ever wondered about April showers? They bring May flowers, right? True enough. But it's not quite so one sided as all that. But, I'm getting ahead of myself. Let's start at the beginning.

Near the beginning there was water. The same amount of water as there is today, about 326 million cubic miles of the stuff. An appreciable amount when you consider that a cubic mile of water equals better than a million million gallons. Billions of years ago, however, earth's water—made up of two hydrogen molecules and one oxygen molecule, the familiar H20—was trapped in rocks in the earth's crust. It seeped out, excruciatingly slowly, and began the great chain of recycling we call the water cycle. You remember: The sun's heat evaporates ocean water, which rises as vapor, only to be cooled and fall back to the earth in the form of precipitation. That is why the amount of water is still the same today as then. So what about those May flowers? Hold on. We're getting there.

The seeming trouble with this whole water system is that while the amount is the same, the distribution isn't. They heaviest rainfall in the world is on the southern slope of the Himalayan range in India. It gets up to six hundred inches a year. Mind you, the northern slopes get only nine inches a year, about as much as our driest state, Nevada, but still a lot better than Arica in Northern Chile. It has an annual rainfall of two-hundredths of an inch. But lest you think that there's no justice in the world, consider the fact that if the earth's precipitation were evened out, the land portions of our planet would get over two feet of rain a year. If it all fell at once, the earth would be covered a yard deep.

I don't know about you, but I'm not quite ready for that. Neither are my plants and trees. We all live here for a reason.

The word for it is acclimatization, but it just means we have a comfortable fit. The basic idea is to take what you need—not too much, not too little, just enough—and then give something back in return. Plants understand that. While they take up water through their roots, they also give it back through their leaves. A single Birch tree gives back about 70 gallons a day. An acre of corn gives 4,000 gallons a day. They give the water back as vapor, which then rises, condenses, and falls back as rain. And those April showers and May flowers? They sort of cause each other. So, while I wouldn't say that I have learned much about life from my plants, they surely have set up a good example to follow: Like the saying goes, you get what you give.

Have you ever wondered about life? No, not those big questions like "how did life on earth begin?" or, "where is human life headed?" It seems to me that those kids of questions, while interesting, are the mental equivalent of ellipses, those three dots that unimaginative writers sometimes use to end a sentence: "But, he opened the door on his left instead…" Like a story without an ending, questions which cannot be answered allow us to speculate endlessly, but to what purpose I wonder? This may be bread and butter to a philosopher, but I get more out of the day-to-day business of living than I do from making tight circles in my head pondering life's imponderables.

Even sitting at my desk I find it difficult to shut out the life around me in favor of what is going on exclusively inside my mind. At this moment, for example, in the bougainvillea outside my window a jaybird family is busily chattering away. It's hard to call any of their talk "singing". Annoying is closer to the truth, and that is quite likely what it is meant to be. The older bird is directing her squawks at me, probably a warning to stay away, while the youngsters in the nest are worrying their parent for food. As soon as I back away from the windowsill she will leave and return shortly with a beak full of bugs for them. And with good reason.

Their annoyance with my intrusion into their world is understandable when you consider that a baby bird will eat its' own weight in food each day. How would you like to have to satisfy that sort of an appetite? And it doesn't exactly stop with adulthood, either. A mature bird will eat around one-twentieth its' weight every day. In perspective, if you or I were to "eat like a bird", we would have to consume three to four times the amount of food we generally take in daily.

Birds, you see, have to catch quite a few insects each day to survive. And considering how prolific insects are, I am eternally grateful for the appetite of my young jaybird neighbors. Dr. Donald Borror, Professor of Entomology at Ohio State University, has for instance calculated that if all the descendents of a single pair of pomace flies were to reproduce unchecked for one year, they would fill a ball 96 million miles in diameter, more than the distance from here to the sun. Even a very hungry jaybird family would have trouble keeping up with them. Thankfully, however, our bird to bug ratio is self-regulating. We humans really needn't do anything except stand by and watch. So, I don't really think much about where human life is headed. The balancing act nature is putting on outside my window right now is too good to miss a moment of.

Have you ever wondered about D-Day? To be precise, we should talk about "D-Days." The term itself came about during World War II, when the Allies used D-Day to designate the date on which troops would land on enemy held coasts. This year, of course, marks the fiftieth anniversary of the most famous D-Day of them all, the allied invasion of Normandy. But, that was half a century ago. A lifetime to some people and more than that to those of us born at the time of the Korean war or the Vietnam war. For us, that war, that landing, that day is not a memory but rather an impression gained from history books and movies.

For me, it's been a very lasting impression. I'm talking big screen Panavision time here. You know the feeling. The theater goes dark, the overture slowly rises, till, in mile high letters, the title of the feature blazes across the screen: THE LONGEST DAY. I first saw that movie in high school and I am a bit embarrassed to admit that for years afterwards I thought that June 6, D-Day, was actually the longest day of the year. Wrong. Just movie hyperbole. In truth, that honor generally goes to June 21. At our latitude, the longest day has around 14 hours of daylight, as opposed to the shortest day, usually December 21, with only some nine hours of sunlight. And the vital factor is light from the sun.

The lengths of the days change, you remember, because the earth is tilted slightly. As it circles the sun on its yearlong orbit, first one hemisphere, then the other is likewise tilted towards our star. Hence while Christmas falls in our wintertime, it comes in high summer for Australians. The process takes time, of course, and as we slowly tilt and turn our way through spring towards summer, the days lengthen with our increased exposure to the sun's light until we reach the pinnacle of the longest day and from there begin the gentle slide back down into autumn. A peaceful routine, isn't it? Well, try speeding it up. A lot. No, not the days, the motion of the earth and the sun.

The earth, contrary to our perception, is not tranquilly revolving around a stationary star. Actually, our sun, like very other star in the universe, is moving outwards at approximately 43,000 miles an hour or twelve miles in a second. Fortunately for us, its' enormous gravitational pull is also pulling our entire solar system along with it at the same rate of speed. So, I guess the history books and the movies are sort of right. We have come a long way since June 6, 1944; about 19 billion miles. But, considering the wars going on today, I do wonder, where are we going so fast?

Have you ever wondered about the Fourth of July? Just what are we celebrating with our fireworks? Our nations' birthday? The Declaration of Independence? Are you sure? How much do we as a people know about it?

Well, it was signed on the Fourth of July, right? Not quite. Actually, our declaration of independence from England was adopted on July 4, 1776, but the document now preserved in a helium-filled case in Washington, D.C. wasn't part of that event. It was ordered afterwards, the original wasn't on parchment, and was signed on August 2 by those members of Congress present. Everyone else signed later.

Thomas Jefferson wrote it, didn't he? Yes and no. He did write it, but there is little in our Declaration of Independence that was new. Ironically, the truths, which we hold to be self-evident were English in origin, expounded most fully by John Locke in his 1688 work, "Two Treatises of Government."

But, did the Liberty Bell crack when the declaration was being read publicly for the first time? Sorry, it cracked for the first time in 1752 and was repaired. It cracked beyond repair in 1846 in celebration of George Washington's birthday.

So, just what are celebrating this Fourth of July? Not myths surely, but rather the reality of our national belief in the equality of all men and their rights of life, liberty and the pursuit of happiness. But that rings rather hollow when you consider that, because of slavery, African Americans could not vote until 1870. Women waited until 1920. Native Americans? They got the vote in 1924. Are we then celebrating an ideal as shaky as our historical knowledge? Not at all.

You see, by 1787 it had become clear that the United States needed a better operating plan than the Articles of Confederation. Out of that necessity arose our constitution, the quiet fulfillment of the promises made a decade before. But not without a fight.

The Constitution was first proposed by the elite of our new nation. With such parentage it was viewed suspiciously by the common people who feared that a strong federal government might create a new monopoly of power. They called for guarantees of such basic human rights as freedom of speech, of print, of assembly, and of religion. Still and all, the Constitution passed on July 26, 1788 by a vote of 30 to 27. The Bill of Rights followed in 1791. And therein lies the essence of what we are celebrating this Fourth of July. The ability of we the people to grow. And grow we have. The 26 amendments to our nation's charter are proof of that. So, happy birthday to us, and many more to grow on.

Have you ever wondered about termites? As pests they haven't a peer. You can swat flies, repel mosquitoes, spray hornets, and catch pesky mammalian pests like mice, but what can you do about something you don't generally see? Not much really.

Termites, you see, are wonderfully well adapted to our lifestyle. Even though they prefer to live below ground, out of the elements, they are not above visiting their upstairs neighbors for a meal. But don't think them ungracious, they never eat and run. By constructing mud tunnels from their subterranean nests to the object of their desire, they can come and go with relative ease and stay as long as they like. And they like to stay a long time.

Termites are among the longest lived of all insects; one captive king and queen lived for 25 years in an observation nest surrounded by their offspring, ten million of them over the queen's lifetime. And they're born hungry. It is, of course, their hunger which makes them such a pest because it is generally satisfied by wood. But not wood per-se. More specifically, termites crave cellulose, the inert carbohydrate which forms the major part of a plant's cell walls. So, as soon as these babies are born they cry for cellulose and during the first twenty-four hours of their lives they get it in the form of predigested food given to them by the older termites. In this termite pap are myriad tiny one-celled animals called protozoa that will then break down the cellulose into digestible substances. Without these little riders, a termite could eat all day only to starve to death. And good riddance, you say. Except for their protozoan hitchhikers, who needs them? Well, we do.

They are part of nature's food chain. And they enrich our soil by decomposing plant material, but you know, I really can't bring myself to like termites much. Yet, isn't that just me showing off my prejudicial ignorance? John Muir wrote in 1867 "The universe would be incomplete without man; but it would also be incomplete without the smallest transmicroscopic creature that dwells beyond our conceitful eyes and knowledge." Creatures such as the tiny mites that live their whole lives in front of "our conceitful eyes," six to a hair at the base of our eyelashes. These follicle mites feed on our cellular fluid and appear to be quite benign, so why are they there? Honestly, I don't know.

For that matter, what about their still tinier cousins, the millions of dust mites which live in our beds, carpeting and furniture and eat the millions of dead skin cells we lose each day? Could we live without them? I wonder.

But, I also wonder, which of us is wise enough to say for sure? Which of us is brave enough to try? Not me.

Have you ever wondered about soil? Call it what you will; dirt, land, property, country, earth, it has been of vital importance to our history for as long as records have been kept and probably before that. Families have been split over it, wars have been fought over it, and always, lives have been lost.

Consider the great peasant revolt of 1358. In that fateful year the farm laborers of France quite literally turned their plowshares into swords and rose up en masse against their landlords. Under the feudal system, a commoner was born, lived and died on his lord's property, with little real chance of ever owning his own his plot of land and thus rising above his born station in life. But these peasants rallied to the truly revolutionary observation that the soil was originally intended for those who worked it, not for a few privileged owners. They pointedly asked, "When Adam delved (farmed) and Eve span (made cloth)— Who then was the gentleman?"

The revolt, however, failed. The peasant's plowshare swords proved no match for the legal pens of the landed aristocracy, but the seed was planted and today, 600 years later, France and the rest of the world is a quite different place. For one thing, the common man is now generally able to own his plot of land. Not a small feat when you consider that while the world's population has grown considerably since the fourteenth century, our land mass hasn't grown larger to accommodate us. Or has it?

Soil, you see, is a rather precious and hard won commodity. Our peasants could have told you that, but the story is deeper than they might have realized. To a scientist, soil is the end result of weathering, the combined efforts of winds, rain, ice and sun, and the largely invisible work of soil fungi, bacteria and invertebrate animals such as earthworms. Imagine a layer of bedrock being broken apart into smaller and smaller particles by the process of weathering. Carry it to an extreme and you have a dry sandy beach, but add microscopic bits of clay and silt and you can have water retention to go along with the oxygen held in place by your sand. Throw in some organic matter like store bought potting soil, in nature it is called humus, to balance aeration and moisture levels and you can begin tilling your soil in earnest. Sounds simple enough, but, like most simple things it is easier begun than finished. It takes nature, you see, approximately 300 years to produce one inch of good soil. In the six centuries since our peasant friends set the stage for us to own our own plots of land, we've only gained about two inches of earth. So, while Adam and his children have had to till the soil since Eden, it could have been worse. We might have been required to make it.

Have you ever wondered about bats? Not as much as you have about dogs or cats I would venture to say. Bats just don't evoke in us the warm, cuddly feelings we associate with our pets or the tolerant camaraderie we have towards domesticated animals such as bees. Instead, these Halloween habituals tend to frighten us, but why?

Contrary to popular opinion, these common animals—bats comprise one-fourth of all known mammals—are really rather benign creatures. Fewer than 0.5 percent of them carry rabies. In fact, according to Dr. Gary Graham, a world authority on bats, the total number of deaths caused by rabid bats since the Korean War is less than that of dog bites and bee stings for a single year. What about vampire bats? Being a bat's meal can't be healthy! True. It isn't particularly good for anyone but the bat, but your chances of being bitten are pretty slim. Check out the figures.

Of the 1,000 bat species worldwide, California's 24 species represent the second highest concentration in the 50 states. Only Arizona has more, with 28. But, of our 24, none eat blood. Nor do our Arizonan neighbors. Fact is, only three species, all south of the Mexican border, subsist on blood—about one ounce per night per bat—and only two of them seek mammalian blood, the third preys exclusively on birds. None actually suck blood. Instead, they make a small incision with their fangs and then lap the blood as a cat laps up spilled milk. Gross perhaps, but there is scientific hope that the saliva of these vampire bats, which stops blood from clotting, may also help alleviate blood clots in humans and prevent heart attacks.

The great majority of bats, some 70 percent, eat insects only; on average about one-half their body weight each night. To put that in perspective, there is a single colony in Texas that eats some 500,000 pounds of insect pests each night.

So what about the 30 odd percent? Well, they're primarily plant eaters and subsist on fruit flesh, fruit nectar, pollen and leaves. In North American, squirrels, birds and other small animals create bio-diversity by distributing seeds while bees and other insects pollinate flower bearing plants and trees. In many other parts of the world, most notably in our rapidly dwindling rain forest, bats serve the same useful purposes. And not just in the wild. Bats are largely responsible for crop pollination is many parts of the middle and far-east. In fact the Chinese word for bat, "fu", also means good fortune. So maybe bats aren't as bad as we think they are. Only, one last thing. They're not flying mice. Their closest relatives are primates, not rodents. Primates, like you and me. Just a final thought for Halloween.

Have you ever wondered about Thanksgiving? Traditionally, it runs a close second to Christmas, but any airline will tell you that there is no real contest; everyone wants to be home for Thanksgiving. Why? Silly question. Because for most Americans it is the time of the year when we put aside our differences and come together to give thanks for our bounty; just as the first pilgrims did with their Indian allies on that first Thanksgiving some three centuries ago.

Imagine how thankful those pilgrims must have been just to have been alive after that frightful first winter when so many died. But what of the Indians? As you remember, the pilgrims were helped immeasurably by an Indian named Squanto, but what is not generally remembered is that Squanto, whose real name was Tisquantum, had only recently returned to the New World. He had been kidnapped in 1615 by an English sea captain, sold into slavery, taken to England and then returned to his homeland in 1620, only to find that a smallpox epidemic had virtually decimated his tribe, the Wampanoag. Thanksgiving was, however, offered by the churchly colonists for the deliverance to them of the new world through the three year long plague that providentially killed 9 out of 10 Indians, "chiefly young men and children, the very seeds of increase" as one Puritan minister put the matter in a sermon he entitled, "The Wonderful Preparation the Lord Christ by His Providence Wrought for His People's Abode in the Western World." The Puritans, you see, labeled themselves God's chosen people.

But, that was long ago, and far away. Sadly, the history of our state, of our area, is neither. Los Angeles, far south of the fabled gold fields, didn't participate in the rush of 1849, but here there was a different kind of wealth; land. Trouble was that the land was already inhabited by various indigenous tribes. The solution was as simple as the logic that spawned it. A government inspector noted in 1858 that the "great cause of civilization…in the natural course of things, must exterminate Indians." And so the preparation continued, still with a thankful nod for the bounty thus procured. The settlers, you see, labeled themselves God's chosen people.

But, though that was really not so long ago, and not far away at all, such logic and actions seem worlds apart from us and so distant that we cannot even imagine a common ground. Still and all, it exists in our holiday. So I'll give thanks for my bounty, but I'll also pray for Bosnia, Haiti, Rwanda and those places in the world, which are not so far away and the deaths there, which are sadly not so long ago. Because, you see, we can no longer afford to be choosy. Else, who is next?

Have you ever wondered about mistletoe? We all recognize the small green sprig suspended overhead during the Christmas season, but what is it? A plant? Yes, but a special type of plant. Mistletoe is a parasitic shrub about one to two feet across. That's right, that tiny piece you bought in the cellophane bag is just a fraction of the whole plant, but don't get any ideas about quick riches. Mistletoe is not terribly easy to harvest in that it generally only grows high on the trunk and branches of a deciduous host tree, a tree that loses its leaves each year; most often apple trees, but also lime, hawthorn, sycamore, poplar, walnut, locust, fir and oak. Mistletoe, you see, is a bit of a vampire as plants go. It takes the majority of its nutrients from its unwitting host, without whom it could not survive, yet, if the relationship goes unchecked, the mistletoe will eventually kill its provider and thus itself. Rather like a tragic opera, the irony of the relationship is that if it is successful, everyone dies at the end.

Unhappily, trees are not the only possible recipients of this unhealthy love. While the entire plant is considered to be potentially toxic, the translucent white berries of the mistletoe we are most familiar with are without doubt poisonous to human beings. The toxic substances concentrated in mistletoe berries is sufficiently powerful to cause vomiting, diarrhea and abdominal pain in a small child, while a small number ingested can be enough to cause labored breathing, dramatically lowered blood pressure and heart failure in even an adult. To give you some idea of the relative scale of toxicity, the active poisons in mistletoe, toxic amines and toxic lectin, phoratoxin, are among the deadliest poisons known to man. As an example, the toxic lectin contained in the seed of the castor-bean plant, phytotoxin, is some 12,000 times more potent than rattlesnake venom.

So why do we actively cultivate such plants? Just for the sake of such nebulous traditions as stealing a kiss under the mistletoe? No. You see, as with many things in life, it is the use to which the object is put that makes the outcome heroic or tragic. Up North, in Sonoma County, there is quite a thriving mistletoe business going on. Growers up there cultivate and harvest Viscum Album, commonly called European Mistletoe, for sale to European pharmaceutical companies. They, in turn, process those self-same poisons and produce safe dosage medicines that effectively treat mild hypertension and arteriosclerosis in humans. Now, tell me that isn't a Christmas miracle.

Have you ever wondered about our Earth? Of course you have. Who of us hasn't looked at a sunset and thought about the immensity beyond or felt the wind on our face and looked in the direction it came from, as if we expected to see its cause. Who of us has never walked in the rain and felt marvelously alive? I guess it is just that overwhelming sense of aliveness that makes the Earth seem so wondrous to me. No, I'm not suggesting that the Earth itself is alive, though some would argue that case. Rather, it is the incredible diversity of life on Earth, coupled with the tenacity of those myriad lives, their sheer willingness to do whatever it takes to survive and multiply, that makes their home alive. Without its web of life, the Earth is nothing more than a barren rock in a cold, dark expanse of relative emptiness.

The impulse to use the old "oasis in a desert of space" metaphor is pretty strong here, but I am going to resist if. It isn't, you see, very apt. Imagine yourself guiding a visitor to our home. Coming closer and closer from the dark depths of space, the colors of our planet are beautiful, but none are quite so inviting as the blue of our oceans. What our casual tourist would not know, however, is that the salinity of that water makes it unpalatable for a majority of the planet's terrestrial life forms. The water in this oasis seems largely tainted. What fresh water there is begins its life cycle as rain and it is that source upon which most life on earth depends. Thus rain and life are most scarce in the brown areas on the planet which we call deserts.

To our extraterrestrial tourist the blue and brown areas would then seem equally lifeless, neither having usable water available, but we could direct him to come closer and look below the surface of the blue water to see the life within our oceans; animals, plants and fish which rely on salt water whether they breathe air like whales or extract oxygen from the water like fish. With a self-satisfied smile, we agree that our diversity of life is astounding; air breathing creatures on land and in the water, and both fresh water and salt-water fish. Every niche seems filled. But, asks our tourist ward, what about the brown areas? Relax; you're up on this. Our deserts are not dead. There is an abundance of terrestrial life there, but the kicker is that its marshes and springs are equally lively. In fact, the deserts of North America are home to over 100 species of fish, some of which live in water five times as salty as sea-water.

So you see, our earth is quite alive, even in its remote corners. But with apologies to Will Rogers, it has indeed taken a heap of living to make our planet a home.

Have you ever wondered about deserts? To most people the word conjures up images of sand being blown across an arid wasteland of gently undulating dunes, each indistinguishable from its neighbor. Let's adjust that picture a bit and focus on an image of snow blowing across an endless plain of stark whiteness, like in Siberia. That too, you see is a desert, albeit a cold desert. So, just what is a desert? As Mary Austin noted in her classic book, it is a land of little rain. But that doesn't quite tell the story.

In Austin's day, around the turn of this century, a desert was considered quite simply as an expanse of land that received less yearly precipitation than was necessary to sustain human life. The problem with this definition is its annoying exceptions. The Bedouins of the great Sahara desert have managed quite well for centuries and our own Imperial Valley, while undeniably lacking in rainfall, is by no means a desert. The key is not only how much rain falls, but how much remains as usable water.

Today's definition of a desert specifies therefore an area that receives less than ten inches annually. Evaporation, on the other hand, has been calculated at between 70 to 160 inches a year. If water was money, the American West would be insolvent, but so would about one-fifth of the rest of the world. With the single exception of Europe, each of the seven continents can lay claim to a major desert.

North America's deserts form a continental divide nearly as long as the Rocky Mountains, but much more varied in aspect. They begin in southeastern Oregon and southern Idaho with the cold Great Basin Desert. This desert finds its fullest expression in Nevada and Utah where, due to the combined high altitude and northern latitude, more than half of the annual precipitation falls as snow and average annual temperatures are quite low. In southern Nevada the Great Basin Desert blends into our own warmer Mojave Desert, which then transitions into the hot Sonoran Desert of California and Arizona's southern borders. Finally, deep within Mexico, our 300,000 square mile ribbon of aridity ends in the hot Chihuahuan Desert.

But is it all that neat? Well, no. Borders are not exact because of the one variable of available water. The Sahara Desert is growing because of man's demands on the water available on its fringes. On the other hand, the Imperial Valley of California was once desert. Man's irrigation made it fertile. Feast or famine; the choice seems to be ours.

Have you ever wondered about stars? My guess is that it would be a rare person who hasn't at some time in life looked up into the heavens and been awed by the sight. But, like that proverbial iceberg, what we see is only the tip. Let's take just a tiny bit of that tip to start with. Between now and early summer, the predawn sky will be host to a particularly radiant star named Sirius. It's easy to locate, just look to the southern sky any clear morning before the sun rises and you'll see it. Sirius is located in the constellation Canis Major, or Great Dog, so since the time of the Romans, it has been known as the Dog Star. Next to our sun, Sirius is the brightest star that we can see without the aid of telescopes. That puts it in a fairly exclusive club in that there are only about 2,500 stars than can make that claim. And the most exclusive star of that club? For us it would have to be our own sun, Sol. But, are we being unduly prejudiced?

As important as our star is to us, it cannot make any claim to being other than ordinary, albeit in the spectacular way of stars. Sirius, for instance, is three times the size of our sun, but it doesn't even approach the immensity of such super-giants as Betelgeuse—the real star of the movie *Beetlejuice*. Betelgeuse, to the north of Sirius, is 300 times the size of Sol, about 260 million miles across. To put that in perspective, our sun could contain one million earths. Betelgeuse would hold three hundred million. But we're not done yet.

There are still the super-giants such as Alpha Herculis. This star is some two-and-a-half billion miles across; twenty five times the distance between Earth and Sol. Difficult as that might be to conceive, and remember, it's about the distance from the Earth to Neptune, it seems like an evening walk when we start computing the distances between stars.

Our host star is 93 million miles away. The next closest star to us is a dim star named Wolf 424. It is 22 trillion miles away. That is 22 followed by 12 zeroes. The farthest star? No one has seen it yet, but to put things in perspective again, the Hubble telescope has photographed stars which are 6 sextillion miles away; 6 followed by 21 zeroes.

Yet, as humbling as such numbers are, there is always the incalculable. How many stars are there? Carl Sagan tried to put it into words with his "billions and billions," but, imagine instead all the grains of sand on all the beaches in the world. That's how many? And of that number, only one particular star is close enough to give our planet life, but not so close as to prevent it. You know, it seems to me that the real incalculable is how wonderfully fortunate we are.

Have you ever wondered what is? I mean have you ever looked around and wondered just what holds everything together? Think about it for a moment. You remember that all matter, be it solid, gas, or liquid is made up of atoms. Discreet little items that are in turn, made up of still smaller parts, maybe ad-infinitum. But that's rather like looking through the wrong end of a telescope. We need to look in the opposite direction to get the truly bigger picture.

Let's start with us. The force every bit of matter exerts on every other bit of matter is called gravity. The bigger and denser the object, the more gravitational pull. That's the reason we can't jump into space, the Earth holds us down. Now, imagine the pull of the sun. It is strong enough to hold our solar system together; all the way out to Pluto, some 5 billion miles away. Look further out now to our galaxy, the Milky Way, with its billions of stars, each exerting a pull on all its sister stars. In its endless spiraling, like a giant pinwheel, the expectation would be that the stars on the outer edges, like a cosmic game of crack-the-whip, would spin off into deep space, but they don't. The reason? Gravitational pull you say? Well, there are a couple of problems with that.

You see, the individual stars are going fast enough to break away from the pull of the galaxy. They have what scientists call escape velocity. The second problem is that the total mass, that is size and weight, of the luminous matter in the galaxies is simply not enough to create the gravitational pull necessary to hold the star clusters together to begin with. So what is holding them together? Scientists call it dark matter.

It's dark because you can't see it or hear it, despite the fact that it is all around you, whistling by you as you read this. So, how do we know it's even there? Think of an incredibly clear lake. You can look through the water to the bottom and see the rocks below, but they are distorted by the water. Now, make that water otherwise imperceptible. You see the distortion, but have no explanation other than something is between you and the object you're looking at.

Scientists do that with the light from stars. That luminous matter is distorted by the far greater amount of dark matter. How much greater? The range is from 1 to 10 percent of the universe being composed of luminous matter, with the remaining 90 to 99 percent being dark matter. In perspective, as luminous matter beings, as a small percentage of a small percentage of what is, we're rather insignificant overall, but I can't help but to appreciate what a great view we have of it all.

Have you ever wondered about pirates? For most of us clothes do make the pirate; gold earrings, heavy cuffed jacket bristling with gold buttons and knives, tricorn hat, cutlass in one hand, flintlock pistol in the other, maybe a wooden leg or a metal hook, certainly an eye patch. Pirates of the Caribbean? You bet. And Hollywood. And Treasure Island. And Peter Pan. And back and back to some dim past. Pirates, you see, really are such stuff as dreams are made of. You think not? Doesn't all play and no work sound like a dream? It was certainly the dream of those who went a pirating, reality notwithstanding.

Let's use the mental image we conjured up a moment ago as a pretty specific example. Though pirates have been around as long as people, the Long John Silver type of pirate we all think of is properly called a buccaneer. That name comes from an Arawak Indian word, boucan; a small fire used to smoke meat. The first rag-tag outcasts to start plundering Spanish ships in the 17th century Caribbean area lived as hunters on the island of Hispaniola, modern day Haiti. They subsisted mostly on the game they killed, cut into strips and dried over boucans. Hence, when the Spaniards took notice of these ever increasing hordes of sea robbers, they identified them by their lifestyle and called them boucanniers, driers of meat. Not a very romantic name, but neither was an ordinary pirate's life particularly inspiring. In fact, it was generally short and brutal. So why did so many join the brotherhood of the coast? Because there existed the opportunity for fortune and glory. And that pervasive dream was universal.

About the time the first boucanniers were experimenting with guerilla piracy in the Caribbean, Kuo Hsing Yeh inherited his father's rather formidable maritime trade monopoly along China's coast. Kuo's father had just been executed by the ruling Manchu court as a 'reward' for delivering the Ming emperor into their hands. The betrayer betrayed, Kuo swore unending vengeance against the Manchu dynasty and fealty to the Ming dynasty. Kuo Hsing Yeh and his pirate navy harassed the Manchu relentlessly until his death in 1683 and is today revered in Taiwan as a national hero. A hero as well as a pirate? A contradiction in terms? That depends. Let's jump forward just a bit in time to someone you know from your school days as a hero, but whom his contemporaries thought of as a pirate; John Paul, not yet Jones.

He was born John Paul in Scotland in 1747. He came to Virginia a teenager and left as a young man, the captain of a West Indies Merchantman, in search of fortune and glory. He came back instead with an untold story and a new name; John Paul Jones. It was under this new identity that he was commissioned a lieutenant in the Continental Navy in 1775 and given a Letter of Marque, a document giving him legal power to harass the enemy as a privateer. Legalities aside, he had been given carte blanche to be a pirate. Sailing to England, he first plotted to kidnap the Earl of Selkirk and hold him for ransom, but the earl was inconveniently away from home and Jones instead took the family silverware. Though he later returned it, the deed was done and, Letter of Marque or not, Jones was branded a pirate by the British. No doubt figuring that birds of a feather should flock together, Jones joined forces with a small band of French privateers and off they set in search of British prey. They did fairly well, capturing three British merchant ships before the small pack was separated and only Jones' ship, the Bonhomme Richard, and a single French privateer, the Pallas, remained together. It was they who came across a British convoy off Flamborough Head being protected by two warships, one of them the Serapis. The Bonhomme Richard and the Serapis squared off and began to pound each other mercilessly. After a time, the captain of the Serapis asked Jones if he was ready to surrender? His reply? You know it: "I have not yet begun to fight!" And he hadn't. The captain of the

Serapis eventually surrendered and Jones won the day. Sort of. You see, while his bravery made him a hero in France, his actions made him a pirate in England. America? We didn't know quite what to do with him, so we did nothing. Jones eventually would up in the Russian navy, but found no peace there; he was accused of assaulting an under-age girl. Guilty or not, he fled to Paris and died there, destitute and forgotten, in 1792. It would be nearly a century before his remains left their unmarked pirate's grave and came home to a hero's monument, but I wonder, which really suited him best?

Have you ever wondered about change? If you have children you certainly have. But even if you aren't in especially close contact with the next generation, you are definitely caught up in their interest. Consider how many times this past month you've heard, read, or seen something that concerns the concept of "morphing"; as in really powerful morphin rangers. As any child can tell you, to morph is to change shape. Perhaps not the best use of a very old word, morph is ancient Greek and roughly translates to "form", but, by changing from a noun to a verb, it is at least an excellent example of how malleable language is. Everyone knows how words can be twisted to persuade, or distorted to confuse. How many times have you reminded someone else or been reminded yourself that there are two sides to every argument? If you have children, more times than you can count. But, are there necessarily two sides to everything? If you are a mathematician, no.

There is, you see, a special type of geometry that is called topology, the mathematics of distortion. It allows us to study figures and surfaces when they are distorted, morphed if you will, from one shape to another. Take a circle, for instance, it has only two sides: inside and outside. Stretch it in four directions and you can make a square, a rectangle, or a quadrilateral. You've morphed your circle. So what, you say, it still has an inside and an outside. True, but it also has four other sides it didn't have before. Stretch it in five directions and you have a pentagon and five new sides, and so on. We're on safe ground so far. Let's get dangerous.

Take a strip of paper about a foot long and two inches wide and make a ring by taping the ends together. There's your circle. You can physically stretch it to make all those figures we talked about. You can morph it all you like, but it still has two sides to it. Next, cut an identical strip, but before you tape this one together, make a half-twist in it.

Now, get two different colored crayons or markers and your first ring. Take one crayon and draw a line on its outside without lifting your crayon. Do the same on the inside. Two sides, two colors. Try the same on your second, twisted ring—remember, don't lift your crayon! You'll never get past the first color. Your second ring only has one side. Its called a Mobius Strip, after its nineteenth century discoverer.

Interesting, you say, but what good is it? Well, because of its unique electrical properties, the U.S. Atomic Energy Commission has patented a nonreactive resistor band made like your ring, albeit out of metal. But, one last thing; a word of caution to parents. Don't waste this on your kids. They already know there is really only one side to any argument.

Have you ever wondered about energy? It was a major topic back in the seventies; remember gas lines? And how about alternative energy sources like wind, geothermal, and solar? They seem to have taken a back seat in the public's mind, but they're still around. Take solar energy for instance, it's everywhere. And that commonness is part of the sun's public perception problem. We tend to take it for granted.

Ask most people what solar energy is and they'll give you an example and not a definition: It's a solar panel. We've all seen them and a lot of us have them to heat our water, but equating a solar panel with solar energy is rather like comparing a light bulb to light. In fact it is released energy that heats our water or lights our rooms. Energy, in the scientific sense is therefore the ability to do work. The more energy available, the more work that can get done. And what energy is more readily available than sunlight?

As far as human beings are concerned, it has always been around. At about 5 billion years old, the sun has always shined on our few thousand years of recorded history and has in its lifetime pretty much always given the earth the same amount of heat and light. If the sun had been a lump of coal it would have burned out after only some 1500 years. The sun, on the other hand, gives off enough energy even now that if we could figure out how to harness it all, every square yard of sunlight could give us two horsepower of energy. But that's peanuts compared to the source itself.

The sun, you see, has a very common, but spectacular way of making its own energy. It's common because every other luminous star in the universe does the same thing. It's spectacular because of the show it puts on while it does it. To make its own energy, the sun merely fuses its hydrogen atoms together to make helium atoms at a ratio of four hydrogen atoms to one helium atom. Think of it as an atomic bomb in reverse. In this case the atoms are not split, they are fused together, but with the end product, helium, weighing 1% less than the hydrogen that came together to make it. What happened to the lost mass? It turned into heat and light energy.

For every 400 pounds of hydrogen changed into helium, three pounds of the sun's matter becomes heat and light energy. That might not sound like much, but in fact, the sun is changing its matter into energy so fast that it loses weight at the rate of four million tons a second. But, don't worry, even at that rate it can send out an undiminished amount of energy for another 35 billion years or so. Maybe by then I'll have put out enough energy to have worked off that 10 pounds I put on at Christmas.

Have you ever wondered about the power of the press? Like most things in life, good or bad, the so called fourth estate has nearly always been with us, but it has seemingly taken on a more sinister cast in these troubled times of ours. No, I am not talking about the rhetoric, which is so rife these days, condemning the press for what is labeled its liberal bias. Rather, it is that whether we like it or not, we are now living in the fast lane. In some capacity we are all on what is being called the information superhighway and when we watch or read our daily news those Internet headlights all too often seem to be focused not on our future, but rather uncomfortably on us. In that glare, an issue like political bias seems to me to strike rather a feeble flame.

Consider. How many times in the past year have you read about the dangers inherent in an uncensored press? How many times have you read about the dangers of the uncensored Internet? It would seem that one's view on Freedom of the Press depends to a great extent on which press you subscribe to: print or electronic; and which philosophy you ascribe to: let-it-be *laissez faire* or let-it-rip *laisser-aller*. But, who can say with finality, which is truly better?

The framers of our constitution knew that without restraints, freedom of the press would result not in real freedom, but in anarchy. If everyone were allowed to print whatever he pleased, regardless of its impact on others, individuals would often be prevented from making or living by their own principal choices. For instance, it is safe to say that most of us favor security. We place a high value on our safety and that of our families and our country. To that end we have a justice system and laws, which exist to protect us. It is something, which we, as

American citizens, have come to prize and expect. But what about that pesky Freedom of the Press? What about the freely given opportunity to put out for public dissemination practically anything? How about the much-discussed availability of specific instructions for making bombs? It certainly gets a little bit trickier then. We should be both personally free and personally safe, but where do we draw the line?

In 1995, anyone can learn how to make homemade bombs through the Internet. In 1775, 60,000 American colonists learned how to make gunpowder in the pages of Nathaniel Ames, Jr.'s almanac for that year. Without that information, without that power of the press we might not even be considering these fine points today or celebrating our fine freedoms this Fourth of July. No one ever said freedom for all was that easy.

Have you ever wondered about the moon? To the ancients the moon was a god or a goddess. They believed in the power of celestial bodies to influence the lives of human beings and so personified them. The Romans named her Diana, after the goddess of the hunt. Just look at the next crescent moon and you'll see her bow as clearly as they did, as clearly as any child today can show you the "man in the moon." And before you smile at their innocence, consider that the child was likely taught that bit of lore by an adult. For us today, despite our perceived sophistication, the moon is thus as close to us as it was to the firstborn of our species. To my mind, that sense of continuity, of order, is a good thing. Better at any rate than its counterpart; chaos.

Now you and I probably share an idea of what chaos is. Utter confusion or disorder. Right? But, apply it to the rest of the universe. To a scientist of today as well as to an Egyptian priest of 5,000 years ago the concept of "chaos" would be the same; the infinity of space or formless matter that preceded the ordered universe we live in. Newton likened our universe to a great clock wound up by God and left to tick on till the end of time in perfect synchronization. To a modern scientist, chaos is the upsetting of that balance.

What if, for instance, the earth moved the thickness of a dime around its solar orbit? At present, the earth occupies the thin zone around our star in which water remains liquid. Move the earth that dime's width and you start us moving either closer to the sun or farther away. Eventually our oceans would either boil or freeze. Luckily, that has never happened.

What about the ice age? Well, they were caused by a slight change in the tilt of our earth's axis of rotation. Remember, the earth is tilted about 24 degrees. Make it 25 degrees and you have enough climactic instability to cause an ice age. Luckily, that doesn't happen often. But why? Because of the influence of our moon.

Our moon, you see, exerts enough of a pull on the earth to keep it at an even tilt. Planets such as Mars, with inadequate lunar pull, actually wobble on their axes, and thus haven't the climactic stability necessary for life. The upshot is that to sustain life, a planet must first be within its star's habitable zone. Second, it must have a moon large enough to steady its rotation. And how likely is that to happen regularly? Even in an ordered universe, not very. The moon may not only be our nearest neighbor, it may also be our oldest and closest friend in what might after all turn out to be a rather cold and lifeless universe.

Have you ever wondered about cats? If you have, you're probably fond of them. You might even be in the category of "cat lover." If you are, you know already what I'm talking about, but if you aren't, well, take a look at a cat sometime. I mean a good, honest look. Set aside your prejudices and you'll see that cats look at you through eyes that speak eloquently as any movie star's, and without having had acting lessons! If you're still not sure, however, what that cat is saying to you, allow me to translate: I'm an old hand at cats.

If he looks at you with his eyes wide open, it means, "Everything I see is mine." If his eyes are half closed, it means, "I know something you don't know…but I'm not telling." Now, it's that second look that has traditionally gotten cats into trouble. It's true that the ancient Egyptians thought enough of cats to make them sacred and to mummify favorite cats so they could accompany their preferred humans into the next life, but by the time the dark ages rolled over Europe a thousand years later, cats, because of their secretive ways and looks, were generally thought to be in league not with the gods, but with the devil. If cats have nothing to hide, the question went, why do they act as if they do?

Like our cat, Meatloaf. He quite literally dropped in on us three years ago. Failing to get his intended bird, he fell out of a palm tree in our back yard and was so stunned on hitting the ground that he looked like a little grey meatloaf. Since that time he has reigned exclusively from indoors, his throne whatever windowsill affords him the best view of his outer demesne. The best seat in the house, however, is still the garden-window in the kitchen. When we had it put in I had visions of using it to grow herbs and paper whites and such, but his majesty voted down my plans by the simple expedient of eating whatever plants I put in his path. So, loyal and understanding subject that I am, I quit trying. Until this summer. Though I hadn't planted them, I found periwinkles growing in our back yard. Soon we had so many periwinkles that I could steal one away for the garden-window. Where Meatloaf promptly ate it. I was angry, of course, but he just looked at me with his half-closed eyes and I gave up again.

Now weeks later, my periwinkles are starting to die off for the coming cold times, but they'll be back. Not like some other plants. You see, each and every day, while we humans worry about spotted owls and such, five species of plants become extinct. Does it matter? Well, periwinkles, for instance, contain substances used to combat some 13 forms of cancer, including leukemia; a disease cats can get. And so I wonder, does Meatloaf know even more than we do?

Have you ever wondered about spiders? Not your favorite subject, right? Well, you're not alone. By inexact count, about one in four of us has a deep-rooted fear of something. Psychologists call it a phobia. One of the more common phobias is arachnophobia, fear of spiders. It isn't, however, a new fear. Our scientific name for spiders, arachnid, comes from Greek mythology. Arachne was an arrogant Lydian woman who challenged the goddess Athena to a weaving contest. Arachne lost and as punishment was turned into a creature of dark corners and caves; a creature which people would fear, the first spider. Athena did, however, show compassion. She allowed Arachne to retain her skills, albeit as a weaver of webs.

In truth, of course, spiders go back much farther than the Greeks. By today's estimates, the family of man has been around in some form for about 10 million years. By the time human beings got the bright idea of going into caves to get out of the rain, round about one million years ago, the family of spiders had been living in caves and just about everywhere else for nearly 500 million years. At present, taxonomists list 30,000 species inhabiting all areas of our planet except the air. Spiders are quite adaptable, you see, but what made them so adaptable? Mostly their fine ability to turn water soluble protein into insoluble silk lines so fine that ten of them would equal the thickness of a human hair, yet be stronger than a steel line the same diameter. Not that the spider knows that or really cares. For her, it's just a way to catch food; for us, it's a little miracle. And a darned useful one.

For instance, Black Widow silk was used during WW II to make cross hairs for bombsights. It worked so well that an optical firm in

Pittsburgh still uses spider silk for their surveying instruments, but let's go past the present use of old technology and take a short leap into the future.

How about a surgeon using spider silk micro sutures? How about a protective coating of spider silk for the space shuttle? Sound far-fetched? It isn't. At present, the University of Wyoming and the U.S. Army are researching these possibilities and others. Their only trouble is getting enough spider silk. Spiders are very territorial, so you can't breed them. The solution? Clone the genes which spiders use to make their silk and make your own. Injected into E. coli bacteria, these cloned genes have reproduced spider silk proteins. But, the process is slow and at present not substantial; no actual fiber has yet been produced, only protein globules, but it is a start and patience is a virtue. Ask any spider.

Have you ever wondered what life is? At its most basic level now. No fair discussing quality, that's a value judgement, or purpose, that's philosophical. Just the simple question; what constitutes life? Where to begin? Let's use an old game: Animal, Mineral, Vegetable.

First and foremost, to qualify as a life form, the thing in question must be able to grow. People grow up. There's animal. Stalagmites grow upwards. There's mineral. Trees grow upwards. There's vegetable. So, we have all three kingdoms covered, but with a slight flaw. Let's qualify that requirement to read, "grow of their own accord." In other words, they have a metabolism that allows them to grow. Sorry, stalagmites. Minerals lose on the first round because while they grow, they can't metabolize nutrients to do so.

Rule two is that the organism must be able to adapt to its environment. It's easy enough to see that in humans. We have managed to live in every known habitat in the world, no matter how inhospitable. Have you ever been in Las Vegas in the summer? How about the vegetable kingdom? Equally adaptable. Moss and lichen can weather the coldest winters imaginable while living on rocks in Greenland. Not for me thanks.

The third requirement is that the organism be able to reproduce. Again, humans have that one beat all to heck. If the Earth were one huge McDonald's, the sign would read "Over 5 Billion Served." And that doesn't include former customers. They've developed a number of ways to reproduce themselves; spores, seeds, runners, and so on.

So, we're left with the animal and vegetable kingdoms. They have life. Obviously. Look around you. What do you see besides animals and vegetables? Just ignore those pesky rocks, they're sore losers. But, let's look a bit closer.

Consider bacteria. They're so small that it would take 1,300 of them, hand in hand, to cross the head of a pin, but they easily qualify for membership in the life club. Good thing, because we larger members of the animal kingdom and our allies in the vegetable kingdom need these little guys to help us metabolize, adapt and have the chance to reproduce.

You see, bacteria are like little chemical factories. They "eat" by breaking down organic material. Escherichia Coli, good old E. coli bacteria live in our intestines and chemically remove otherwise harmful material. Nitrogen fixing bacteria give plants the nitrogen they need to grow and metabolize and reproduce. No bacteria would mean no vegetables, no animals, no life—no joke. Just rocks.

Have you ever wondered how small small actually is? For most of us, fleas come pretty darn close to the limits of our concept of small. It's not that we don't know there are smaller things out there, but it's hard to get a grasp on something we cannot see and fleas are at about the outside limit of most people's eyesight. But, as Ogden Nash noted, even big fleas have little fleas that bite 'em, and so on ad-infinitum. But let's start with the big picture.

How many different life forms can you name? Be specific now, no big groups allowed. How many do you think there are? Guess again. No one really knows, you see, but scientific estimates range from 10 million to 100 million. Now, of those uncounted millions, most belong to the animal kingdom; roughly seventy percent. Don't get too uppity, however, as most of these animals, eighty percent, are invertebrates, specifically arthropods: insects, spiders, centipedes, crabs, lobsters and the like. Like it or not, we vertebrate animals take up a pretty slim portion of the pie. But not the smallest.

If we get down to species, life forms which share numerous important characteristics, we find that some animals are a species unto themselves and size isn't a factor. Desert pronghorns, what most people call antelopes, are a distinct species of one. So are the anajapygids, an insect found only in California. They're small on two counts. There's only one species and they're only 4 mm long. What's smaller? Bacteria.

A thousand years ago, European scholars hotly debated how many angels could dance on the head of a pin. Then, as now, there isn't any empirical way to prove your estimate, you can't put an angel on a slide, but we do know how many bacteria can span that same pin, around

1,300. Put another way, it would take around 26,000 of one species of bacteria to cross one inch. That's small. But remember what Ogden Nash said.

Despite the size difference, bacteria are subject to some of the same afflictions as we are. Specifically, viruses.

Remember last winter? The flu? You were hosting a virus. A virus is basically a tiny (it would take 10,000 to cross a pin head) reproductive machine with a flaw. They cannot reproduce by themselves; they need a host cell, generally from you or a bacterium. Once "infected" by the virus the affected cell will do the viruses bidding and make more viruses, not in your case, cells, or for the bacterium host, bacteria. That causes infections and illnesses. So what bites viruses? Maybe us. If we can learn how to get viruses to carry beneficial genetic instructions to cells instead of harmful reproductive codes, what couldn't we cure?

Have you ever wondered about the seasons? Looking back on it I think my first exposure to them was on the Howdy-Doody show. Remember Princess Winterspringsummerfall? Kind of equals everything out, lumping them together like that, but the truth of it all isn't quite so neat. The seasons aren't one of our convenient time designations like seconds and hours and weeks and months. Rather they're based on the astronomical position of the earth relative to the sun. Specifically, which hemisphere is tilted towards the sun and which is tilted away. But, you knew that. Let's try another. Which month is hotter in the Northern Hemisphere, July or January? If you said January you weren't here in July. One more. In what month is the earth closest to the sun? July? August? Maybe September? Sorry. The earth is closest to the sun in January. Trust me.

You see, the seasons are not all created equal. Here in the Northern Hemisphere the big winner is summer, with 93.65 days a year. Second place goes to spring with its 92.76 days. Third is autumn at 89.84 days and trailing dead last is winter with 88.99 days. That's a grand total of 365.24 days in a year. Now, that's the precise amount of time it takes the earth to complete one orbit around the sun, a solar year. Every year our 365 day calendar is off by that 24 hundredths extra, or about six hours a year. That's why we have to have a leap year every four. Catch-up year would be more accurate, but when you're as time conscious as we are, it would make people too nervous to be behind all the time.

Why aren't the seasons nice and equal? Because the earth's orbit around the sun isn't. A circle would be easier to deal with, but what it has is an ellipse, an egg shaped orbit with the sun at the bigger end. If you've ever watched a Star Trek episode you know that the closer you get to the sun, the greater the gravitational pull and the faster you go; the old slingshot effect. When the earth is at its closest approach, called perihelion, it is also traveling at its fastest rate of speed and so the season in question becomes that much shorter. The season being winter. The month being January. But, if the earth is closest then, why isn't it hot? It is. But not in our hemisphere.

Remember that tilt? In the Northern Hemisphere we are at aphelion, our farthest point from the sun, in July, but we are also tilted towards the sun. So, contrariwise, in the Southern Hemisphere, Australia for instance, Christmas comes in summer. Which for some reason always makes me think of snowballs in you know where.

Have you ever wondered about weight? Yes, I know. A heavy subject. Especially now that the new year is really here. If you're like me the holidays are sort of like a sedentary cruise. Because good food is so available, I eat way too much and by the time I get back to reality sometime in February, my horizons have expanded disproportionately to what I remember from the trip. Think about it. How can such small goodies, even in quantity, result in so much weight?

Well, we can look in a couple of directions. To a nutritionist the correct path would have a signpost that read "CALORIES." That's a familiar term to most of us. It just refers to the amount of energy, specifically heat, which it would take to raise the temperature of a gram of water, about a thimble full, by one degree. Applied to your body, it refers to the amount of energy you can get from food as long as it stays with you. And you know as well as I do how long that second piece of pie stays. That doesn't, however, seem to help us, but keep that thimble of water handy. We'll need it in a moment.

Let's look at weight from a different perspective, a broader one this time. To a scientist weight is the combination of the object's "mass", or size, and gravity. Logically, the smaller the item, the less the mass, the less gravity has to pull on and the less the object weighs. That makes sense. A sugar cube weighs less than a battleship, doesn't it? But, consider that thimble-full of water we talked about earlier. Take a second container the same size and fill it with earth and it will weigh nearly six times as much as the water. The mass is the same, but the weight is different. All right, you say; liquids and solids weigh differently. How about the third type of matter, gas? An equal mass of gas should weigh less than either liquid or solid, right? Well, it depends.

You see, it's not always a simple matter of comparing size and weight. A third container full of steam would weigh practically nothing, but fill it with the gas the sun is made of and it would weigh one and one-half times as much as the water you weighed earlier. Why? In a word, density.

The atoms in our water are not nearly as dense, or compacted, as the atoms in an equal amount of earth, so they weigh differently. But, what about the solar gas? Because of the sun's enormous gravity, it is denser and so weighs more. But consider this. A sugar cube size piece of a black hole, a star which has reached maximum density, would weigh as much as a battleship. I don't even want to consider the calories.

Have you ever wondered about clear? No, you read it right. Clear. Think about it for a moment. What does the word "clear" conjure up for you? I would guess that the image in your head right now has something to do with windows. A view perhaps of a particular landscape you remember? Think of how clear the mountains look early in the morning when the March winds have blown all the smog and haze away. Clear and bright. But, how clear is the image in your head? Really now. Is it as clear as a dream? As clear as a daydream? As clear as reality? And whose reality, yours or mine?

Careful now, we're on a rather slippery slope. Any self-respecting philosopher would have already jumped all over this idea and pinned it to the ground, but I cannot rid myself of the squirmy notion that reality is somewhat relative. Maybe not the big stuff we all see and agree on—the Grand Canyon is certainly grand—but how about the smaller things in life?

The first time my daughter, Megan, saw the Grand Canyon in person she echoed the sentiments of John Wesley Powell and most other early explorers of the region. She said it didn't look real. The sheer vastness of the canyon, its depth, length and breadth, is simply overwhelming, but you still have the sensation that if you could just reach a bit further you could touch the far edge and it would all become comprehensible; it would all fall into place and become real.

The reason behind this confusing phenomenon is clear. Clear air, that is. You see, air-borne particles half the thickness of soap bubble film scatter light like tiny mirrors; think about beams of sunlight coming in your window. When the air is relatively free of particles larger than that, we see more clearly but at the same time we lose some of our depth perception and tend to make silly errors in judgement.

Here's an example closer to home. Let's go back to the San Gabriel Mountains. It's a particularly clear day. In the reality we all share, you know that those mountains are far, far away, but given a chance your mind will trick you into thinking that they are actually very close. Now you tell me, in which of those two instances is your thinking completely free of any obscurities, that is to say, clear?

In neither. It is our shared experiences, knowledge and feelings—consciousness—which allows us to agree on the bigger aspects of reality. But your consciousness is never clear of your personal experience, knowledge, and feelings. It is what makes you—you. Without examining those smaller bits which reflect our personal thoughts and actions like tiny mirrors, how clear can our lives be?

Have you ever wondered about coffee? A lot of people have. So many people, in fact, that serving and selling coffee has become an industry unto itself in the past decade. When is the last time you came across a plain old coffee shop? Now, of course, coffee is a gourmet commodity, but I still go by the old rule of thumb that there are two types of coffee, good and bad. Depending on your taste buds, that's a lot more true than the coffee elite would like to believe.

Coffee, you see, has been around for quite a long time, but there are still only two species of coffee tree; robusta and arabica. Robusta is the poor relation most of us grocery store types drink. Arabica, on the other hand, is the stuff dreams are made of.

No, that's not hyperbole. You see, no one really knows exactly where coffee originated. It might seem clear from the name that Arabia was the source, but Ethiopia might have exported coffee to Arabia first. Popular legend has it that a young Ethiopian goat herder saw his goats behaving unusually spry one day and ate some of the same red berries his herd had been feeding on. Needless to say, he also felt much more awake afterwards. He shared his discovery with the local monks, or so the story goes, and experimentation led to boiling them in water. A nice story, but the truth is, literally, harder to swallow. You see, crushed and dried coffee berries were originally mixed with fat to form food balls, much like the pemmican which American Indians made. Only later was a beverage made from the beans, but it was not what you think, it was a type of wine. Coffee as we know it was first brewed in the 13th century in Arabia, and so Arabica was born.

Coffee reached Turkey in 1554, Italy in 1615, France in 1644 and England in 1650. The first "coffee houses," the ancestors of our upscale coffee shops, were opened that same year in England.

New York's first coffee house opened in 1668, but the story of coffee in the New World really begins in 1723. In that year, a French army officer stole a coffee plant while on leave in Paris and took the plant to Martinique. From the West Indies, coffee eventually reached South America, where it was so successful a crop that today Brazil is the world's largest producer of coffee. We are the world's largest importer. Presently, the United States buys some 25% of the world's annual crop, about one billion pounds. Considering that it takes one tree's total annual output of 4,000 beans to make just one pound of coffee, that's one heck of a job of picking! Is it any wonder coffee breaks were invented?

Have you ever wondered about lawns? Most men have. We have a peculiar gene that compels us to furtively look over at the other guy's lawn to compare his and ours. I'd always suspected that this gene was recessive until the 1950's, when men started to stake out their own little suburban country; but it goes back farther than that.

The concept of a lawn goes back at least to Henry VIII's formal gardens, but the earliest printed mention of the word "lawn" shows up the year after Henry died, 1548, and describes it as a "place void of trees."

More to the modern point of view is a reference in a 1733 dictionary of gardening. That book defines "lawn" as "a spacious plain adjoining to a noble seat." Even then a good lawn reflected the owner's social status. But, let's get on our knees and check out Henry's lawn. See the difference between his and yours? Look closer, his isn't just grass. It's made of chamomile, sanfoin, medick fodder, clover, and a variety of meadow grasses. If you think it looks more like a field, you've partly got the idea. It was nature tamed and contained, but not for any utilitarian purpose like growing food. Rather it was for recreations like croquet, lawn bowling, dancing or just walking. Fields were for commoners; lawns were for nobility because it took real money to make and maintain an unobstructed vista of otherwise useless velvety green grass around your estate. It still does.

Henry VIII knew that as well as you. He just knew it first. What he didn't know was that his stately lawn came from a family every bit as noble as his. Grass, you see, is not just for lords or for looks. The meadow grasses which Henry walked on are part of a family that includes such human food staples as corn, wheat, barley oats, millet sorghum, and rice. Without the grass family, the graminae, you and I and Henry would be sadly undernourished. But this ancient family's value to humans doesn't stop there.

Disease? Have you ever had lemon-grass tea? Or how about those sailors who learned long ago that "scurvy grass" could cure that horrible disease? Comfort? Ever use a mosquito repellent candle in the summer? They're made from citronella grass. Building? Check out bamboo, the largest grass. The list goes on and on, but let's go back home, to our own lawns.

Nothing fancy now, just good old American bred Kentucky bluegrass; a grass whose roots go back to Roman times and whose seeds were likely brought her in the 1600's by French missionary explorers. You know, somehow I can't help but think that my lawn has a nobler heritage than I do. Maybe Henry and I do have something in common.

Have you ever wondered about language? I was out gardening the other day when I heard what I thought was a cat stuck up in my palm tree. I watched carefully and finally figured out that it was a crow making sounds like a cat in distress. Not so amazing when you consider that many birds "talk." But was it talking? Was it a bilingual bird? Or was it just imitating sounds it had heard without really understanding any of the process? I tried communicating with it. It looked interested, but didn't respond and finally just flew away, but I was left wondering when and how we, as higher animals, lost the ability to communicate with the lower animals of our shared world. Doctor Doolittle notwithstanding, it would appear that the answer is more than 4.4 million years ago.

Look at your palm, fingers outstretched. Imagine your palm as the common source for all your fingers, but each finger as a separate and distinct development of humans and other primates. One finger is for human development, one for monkeys, and one for the great apes—they're different families—and so on. Just as your fingers are not attached, so humans did not develop from apes, but go down closer to that common ancestor your palm represents and the differences become slighter. Go back 4.4 million years and you get Australopithecus ramidus, the oldest known hominid, or prehuman. But, how do scientists know it is a hominid and not an ape? Simple, they look in its mouth.

Fossils, you see, are rare, but teeth are very hard and last a long time. By studying the numerous fossil teeth found over the years scientists know that monkey molars have always had four cusps, while human's and ape's molars invariably have five. But, you say, a fossil tooth could then be either from an ape or a hominid. True. It takes one more bit of evidence to make the distinction.

Look in your own mouth and you'll see it. The roof of our mouth, your palate, is curved. Also, your jaw is wider at the back than it is in the front. In contrast, an ape's palate is flat and his jaws are in straight, parallel lines front to back. Find any of these characteristics and you've gone a long way towards identifying your fossil as either hominid or ape.

But, what about language? Well, without that curved palate and the peculiar way our jaws are constructed, complex human type speech is impossible. So, Doctor Doolittle was on the right track when he talked to the animals, because, sadly, they could never talk back in his language.

Have you ever wondered about names? Shakespeare said that a rose by any other name would smell just as sweet and while he was probably right, I still think I could probably detect a difference. I'm guessing you could also. Let's face it, names do affect our perceptions. Consider the Chinese Goose-Berry. For years producemen refused to stock these rather ingloriously named fruits because no one would buy them. They just didn't sound very appetizing. Then someone got the bright idea to re-christen the lowly Chinese Goose-Berry. Today it sells just fine as Kiwi fruit. There's no real difference, of course, it's still the same fruit. Only our perception has changed because of the new name, but doesn't that simple change create a new reality for us? Too philosophical? Well, let's look at it another way.

Remember when the tennis player Andre Agassi proclaimed in a series of advertisements that "image is everything?" My first reaction was to counter that image is ultimately nothing; clothing does not make the man. It is what you do that counts. But, how true is either statement? Deeds should count more than words, more than image, but we do nonetheless tend to define ourselves by the names we give ourselves. Consider our politically correct age in which people are not handicapped, but are "differently challenged." Despite the nomenclature, reality is unchanged, the individual still has a disability, but reality is also changed by the new perception given by the new name. One is no longer helpless, just challenged in ways others are not. Political correctness? Yes. Wrong? I don't know. People smarter than me, the founding fathers, also had a tussle or two with the name problem.

When this country of ours was just getting started, one of the first orders of business was to name it. Two of the biggest vote getters were the United States of Alleghenia and the ever-popular Freedonia. Now mind you, we're talking about Jefferson and Madison and Franklin and Washington, not the Marx Brothers. You have to wonder what they were thinking. Some years later, President Thomas Jefferson saw the need for his new nation to expand itself. Ever the thinker, Jefferson took out pen and paper and carefully drew a map of the existing thirteen states. Then, between the original states and the Mississippi River, Jefferson laid out the boundaries for 14 new states. Two of these new states bore the mellifluous names of Michigania and Metropotamia.

So, be silently thankful this Fourth of July that someone back then had the good sense to stand up and point out that a rose by any other name mightn't smell as sweet after all.

Have you ever wondered about food? If you have ever been to a doctor's or a dentist's office you have surely had time to thumb through any number of magazines and I am betting that at least one of them was a National Geographic. Now stop and consider just how many National Geographics you have flipped through in your lifetime. And how many times have you been stopped cold by that one particular picture?

Now, if you are a baby boomer male you are probably thinking I am referring to another type of picture entirely. Not so. I am speaking here of food, not women in less clothing than red blooded American boys were used to seeing when I was a red-blooded American boy.

And if you are anything like me you were and maybe still are fascinated by what people eat. As a boy I always amazed to see pictures of tribesmen in the jungles of South America roasting bird-eating spiders for their evening meal. The fact that these delectable arachnids were the size of the plate my own dinner sat on was just icing on the gustatory delight of wondering how anyone could eat such a thing. Or how about the nutty tasting grubs which the National Geographic always pictured being dug up out of the ground by Australian aborigines then roasted over an open fire before being consumed like marshmallows round a campfire. Yummy? I don't think so. But fascinating to consider. And I think a human trait.

The Romans, once praised, now vilified for their excesses were no slouches at getting creative with their food. While the masses had to be content with their bread and circuses, the ruling elite knew no such dichotomy. For them food was the show. The one bit of arcane information that has always stuck with me from my National Geographic days was a single course served by a single emperor; peacock tongues. Not the whole peacock, mind you, just the tongues; the rest of the bird being thrown to the myriad citizens who lived by what came their way. I had always thought that was a bit over the top. Surely too precious, a wasteful dainty if nothing else and emblematic of the corruptness of the Roman aristocracy. But, I nonetheless always wondered if those peacock tongues were in fact the best part of the bird? As I likewise prided myself on being a far more socially conscious human than those corrupt Romans, I was certainly not going to find out. I was busy saving whales, how could I ever justify killing a peacock for its tongue? I thought that would be inhuman. And I was right, but that wasn't the whole story.

You see, while I was busy saving whales from anyone except those Inuit who traditionally have hunted and then eaten whales as part of their staple diet, there was a group who have been just as intently hunting whales for a lot longer, and doing so for their own idea of a delicacy, letting the rest go to waste. Humans are not the Blue Whale's only enemy. There is the Orca.

Orcas or "Killer Whales" hunt Blue Whales in packs, kind of like wolves hunt their prey. Usually there are three of them. They begin by circling their huge, but less agile prey, coming from all sides, harrying him to distraction, till finally one of the Orcas will come up sharp and quick from below and bump the Blue Whale full in its belly. Now if you've ever been hit in the stomach you know your first reaction is to open your mouth wide, which is just what the Blue Whale does. At which point the nearest Orca rushes into that huge chasm and takes as big a bite as possible out of the Blue Whale's tongue. And it's a big tongue; about the size of a full-grown African elephant. And just like your tongue, it has a very large vein going through it. A vein which is now severed. It doesn't take long for the Blue Whale to bleed to death, but it takes less time for the Orcas to rush in and begin their own delicate feast. And the Blue Whale? Well, after the Orcas have eaten their fill the carcass will float for awhile, then sink to the bottom where it will be consumed along the way and afterwards by the myriad sea creatures who live on whatever comes their way.

Peacocks, whales, Romans, plebeians, maybe the desire for the new and unusual in food isn't just such a human trait, after all? Something to consider. Or should I hold my tongue?

Have you ever wondered about aluminum? Me neither. Until this morning. Like a lot of people, I have a tendency to chew on my pencils as I am working. From long acquaintance I've become used to the soft cedar wood feel of a pencil, but today's victim, a borrowed pencil, seemed different. It was, you see, an ecologically produced pencil. The "lead" was a mixture of graphite and clay, as is all pencil lead, but was made from graphite that would normally be lost in the manufacturing process. The wood casing was pressed wood pulp and glue. The eraser was man made, not natural latex rubber and the ferrule, the part that holds the eraser to the pencil was made of aluminum. In short, the whole pencil was made of reconstituted materials. Yes, the aluminum ferrule also. Let me explain.

You see, unlike most metals, aluminum is not something that one just digs up out of the ground or smelts from ore. It has to be manufactured from bauxite, the primary source of aluminum's parent compound, alumina. And there's the problem.

It was way back in 1807 that the British chemist Sir Humphry Davy first discovered a new metal, which he named aluminum, in a lump of kaolin clay. He could not, however, separate the metal aluminum from the compound alumina. That honor went to Hans Christian Oersted, a Danish chemist, in 1825. His intense labor produced a few pellets of pure aluminum, the same stuff as my ferrule, but they were so oxidized on the surface that they could not be successfully melted together. It was not until 1855 that the French chemist Henri Deville, after ten years of experimentation, found a way to melt together the tiny particles he made and aluminum became practical, if not cheap. Deville's aluminum cost $545 a pound, a small fortune back then, so it isn't much of a wonder that the first object made of aluminum was a toy rattle for the infant son of the French emperor, Napoleon III. The emperor himself was fascinated by the metal and had a set of aluminum forks and spoons made for himself and honored guests. His less notable guests used the gold and silver tableware.

The problem remained, however, how to make this marvelous metal cheaply. By 1859 Deville had managed to reduce the cost to $17 a pound, but even that was too expensive for every day use. It was not until 1886 that Charles Hall of Ohio and Paul Heroulkt of France, simultaneously and independent of each other developed the inexpensive electrolytic method that makes aluminum so cheap today we can use it for pencil ferrules. I wonder, what would the emperor have thought of that?

Have you ever wondered what the most dangerous animal on earth is? Dangerous to human life, that is. You know where I'm going with this one. For sheer destructive capability, humans surely lead the pack. No other animal fouls its own nest so efficiently as we do. Just as an example, we are cutting down the equivalent of two football fields of rain forest every hour of every day. In the Amazon that works out to nearly 100,000 square miles or about 6 percent of the original forest that we have logged so far. Considering that the Amazon rain forest contains nearly half of the plant and animal species on earth, our wholesale destruction of the forests and the life forms that live there must give us an easy edge. Or does it? Sorry but we're not the biggest danger to ourselves, there are worse things out there. Big cats? I know I wouldn't want to go up against a tiger or a lion. But no, they're relatively scarce. Sharks? Haven't we all seen enough National Geographic specials on the Great White to have developed a healthy respect for them? But, how many of us ever see them up close and personal? Snakes? A better guess. Sea snakes in particular, like the Hawaiian bicolored sea snake, are particularly venomous, but again, how many sea snakes do you encounter on any give day? Anyway, if you really want to talk about venom, try the Australian blue ringed octopus. One nip from a specimen the size of your hand you're dead within minutes. Land snakes? Potentially deadly, but not widespread enough to be dangerous worldwide. Give up? One last hint. They were one of the plagues in Egypt.

If you said flies, you're right. For sheer deadliness the fly family cannot be beaten. That housefly you killed the other day was probably playing host to as many as 6 million disease carrying bacteria at the time of its demise; diseases such as cholera, tuberculosis, polio, plague and dysentery. But even among flies there is a hierarchy. Within the scientific order of flies there is the most dangerous family of all living things, the family Culicidae. You and I are more familiar with their common name, mosquitoes.

Malaria, spread by mosquitoes, kills upwards of a million people a year in Africa alone. But that's not all. These mobile syringes carry around well over 100 different viruses and enough human diseases to make your housefly green with envy. All in all, mosquitoes make some 300 million people a year sick and have killed more people than all the wars in human history. That's quite a record for such a tiny creature. Just something to think about during your Labor Day barbecue.

Have you ever wondered about candy? I'm the first in line aboard the Halloween express each year and I think I always have been. Even as a small boy I looked forward to Halloween with a lot more anticipation than Christmas. I know, I know. It's a pagan holiday. But, let's respectfully put aside the religious aspects for just a moment and focus on the purely material. Halloween meant one thing to me as a child. You got it; candy. Sure, I liked dressing up and having free run of the neighborhood at night, but what would be the point without candy? For that matter, what would life be without candy? To my mind, pretty unsatisfying.

Consider those truly ancient people who left records of their passing only in rock tools or the rarer cave paintings such as those found at Lascaux, France. Candy was unknown to them. The closest they would have gotten to sweets would have been bits of honeycomb stolen from angry bees in an early form of trick and treat. But go forward a few thousand years to ancient Egypt and we find the first recorded recipes for processed confections. We're not exactly talking candy here, more like sweetmeats since honey was still the primary ingredient, but it was a step in the right direction. There was only one prime ingredient missing; processed sugar.

Not that sugar or sugar cane, the primary source of sugar in the ancient world, were unknown. Far from it. Alexander the Great spoke highly of sugar cane in 325 B.C. The Bible lists sugar as a valuable commodity and a physician of the first century wrote glowingly of a "hard honey" which he called saccharum. However, neither sugar nor sugar cane were common.

In the centuries following the fall of Rome, what we call the Middle Ages, sugar was a trade commodity between Asia and Europe, available primarily through Venetian traders who obtained their supplies from Persia—modern Iraq and Iran—Egypt and Syria. Scarcity kept the prices high.

Curious to us, however, with our modern marketing mentalities, is just how long this lack of supply to meet demand lasted. The tide turned in the early 1300's when sugar cane plantations were started in southern France and Italy and by the Spanish in the islands of the West Indies. By the 1400's sugar was more common in Europe, but still rather too expensive for the commoner run of folks, but then, in 1470, a confectioner in Venice found a way to easily refine raw sugar. It was that discovery that began the modern candy industry. For which I am glad. As a boy, I never liked getting apples on Halloween. Getting a chunk of sugar cane at each house would have been worse.

Have you ever wondered about Thanksgiving? I imagine most of us could piece together a pretty warm word-quilt about it. It's a good holiday. One that reminds us about the pilgrims who showed such fortitude that terrible winter of 1621 when half of their number died. Those who survived set a tone of hearty thankfulness for us all to follow.

A century and a half later there again came a time to pause and give thanks for bounties which seemed beyond price. It was the time of the Revolutionary War and eight days were set aside by those who were still whole and alive to give thanks for the victories of our rag-tag army and for the simple gifts of being safe and out of danger.

In March 1789, the citizenry of France overthrew their king and set in motion a revolution that would cause all Europe to feel threatened. In April, George Washington was sworn in as our first president. The revolution in France forced him to make the fateful decision that America would avoid wars in Europe. In June, the Bill of Rights, the first ten amendments to our constitution which seek to make clear our rights as citizens, was introduced into the first United States Congress to meet under our Constitution. Then, on November 26, President Washington proclaimed that the young, but certainly vigorous United States of America would thereafter celebrate a day of thanks for the blessings which we enjoyed in the face of such world-wide tribulations.

In the eye of such storms, many people and many states took to heart the idea that thankfulness needed nurturing. The Protestant Episcopal Church set aside the first Thursday in November, the day of Washington's proclamation, as the day for this thanksgiving, but nei-

ther Washington nor the church called for a national holiday. That
would be left to another.

For some of us, Thanksgiving reminds us of Abraham Lincoln and
that most horrible of wars, our own Civil War, when over 600,000
men died. It was in 1863, after three years of bloodshed with two more
unseen years yet to come, that Lincoln proclaimed Thanksgiving an
annual holiday, a time to give thanks for all the blessings which we as a
nation enjoy.

Thanksgiving serves to remind us of Franklin Delano Roosevelt
who in 1939, on the edge of the Great Depression and the brink of
World War II changed the day of this most national of all our holidays
from the last Thursday of November to the third Thursday in order to
lengthen the Christmas buying season. And there I think we, as a
nation, began to slip. I am thankful for all that I have, but I cannot
help but regret all that we as a nation have lost in our rush to acquire
more.

Have you ever wondered about batteries? Well, what about that Christmas Eve when you suddenly realized that you needed batteries? When you finally found an open store that had them you probably wondered then why something so small and so common as a battery should cost so much. Supply and demand? Maybe partly, but I don't think it's that simple.

You see, it was in Holland in 1663 that the great Dutch anatomist, Jan Swammerdam, began to dissect a frog and discovered that when he grasped a tendon from the frog's leg and touched his scalpel to a nerve in the same leg, the let twitched as if alive. Curious as this must have seemed, Swammerdam never followed up on his "discovery." It would be another one hundred years before anyone else did, and then only because of an accident.

In 1771 an Italian physician by the name of Luigi Galvani laid a freshly dissected frog on a table in order to attend to some more important business. Galvani was at the time engaged in electrical research and had unwittingly placed his frog within the proximity of an electrical machine. As luck would have it, an assistant accidentally touched the frog's legs with a metal scalpel just as the machine sparked. The result was the same as that observed, but ignored by Swammerdam; the leg muscles twitched. The difference was that Galvani began a series of experiments to find out why. In time Galvani discovered that the electrical machine was not necessary if two different types of metal were used. One metal was passed through the spinal cord while the second metal was hooked through the leg muscle. When the metals touched, the muscle contracted. You can try this one yourself. Just hold a clean penny on the top of your tongue and a solid silver coin (like a pre-1964 dime) on the bottom. You'll notice an old taste as the nerves in your tongue record the small electrical current you've created. Just like Galvani, you will have discovered the principle behind the electric battery. But, also like Galvani, you cannot lay claim to having invented the battery. That honor goes to one of Galvani's countrymen.

In 1880, Allesandro Volta discovered that any moist material placed in contact with two different metals would produce electricity. By pairing zinc and copper discs in a small tower or pile and separating the joined pairs with paper discs soaked in brine, Volta could produce electricity of rather low power, but of great quantity. This was the lineal ancestor of those batteries you searched high and low for that Christmas Eve. So, do you still think you paid so much because of supply and demand? Me too. But remember, you and I are paying for 300 years of research and development.

Have you ever wondered about New Year celebrations? By the time you read this you will have already seen the ball drop in Times Square, heard the firecrackers and noise makers make their usual din, watched the calendar pages flutter down into the city streets, toasted the old year and rung in the new and sang a couple verses of Auld Lang Syne. But, did you ever consider how foolish this all must look and sound? Well, there is a pretty good reason. The sister day to our New Years Day, January 1, is April 1. Confused? Well, let's look at April.

In most parts of the world April is the month when the most radical changes take place in nature. The snows of winter are over and the earth begins to green up again. Animals and insects become more evident with the new growth of plants and flowers and people generally seem more active also as outdoor activities call and yard work beckons. As a matter of fact, the very name April comes from the Latin word aprilis, meaning, "to open." Is it any wonder then that up until the 16th century people quite naturally celebrated the beginning of the new year in springtime? What could be more natural? Well, how about the basic human need to encompass our world?

It was way back in 46 B.C. that Julius Caesar set the finishing touches to the solar calendar that bore his name. It was, of course, an immediate success as it was more accurate than any existing lunar calendar, but it was far from precise. A year on the Julian calendar was longer than the actual solar by eleven minutes and 14 seconds. Not a big deal? I agree, but consider that some sixteen hundred years later those accumulated minutes and seconds had thrown the Julian calendar off by ten extra days a year. This was a critical problem for both religious and secular reasons. It was important that religious festivals and feast days be celebrated on the correct day, but it was vital that crops be planted and harvested at the most auspicious times. To do otherwise was to gamble with famine and societal upheaval. No monarch would wish to rule in such interesting times. And for that reason, when a reformed calendar sprung up in 1564, King Charles IX of France adopted it for his country. But traditions die hard.

Until Charles adopted the new calendar, New Years celebrations began on March 25 and ended eight days later, on April 1. When the new New Years Day of January 1 was adopted, traditionalists didn't want to change from their familiar custom. They were therefore branded April fools by those more flexible souls who went along with the change. But you know, I'm not too sure that the real fools weren't those who followed King Charles' lead instead of Mother Nature's.

Have you ever wondered about history? Henry Ford dismissed it as "bunk", but that seems rather too inclusive to me. Yet, on the other hand, can we indiscriminately trust all history?

The first discreet bit of history I can latch onto is the Kennedy assassination. I remember distinctly the moment when the principal of my school came on the loudspeaker and announced that the president had been shot. There was a moment of stunned silence and then murmurs as we tried to understand what she meant. Back then, people didn't get shot. Certainly not presidents. But what was real to me then is now "history." To the generations after me, it has never been anything but history; events to be read about in a book or viewed on a screen. Coming at history without that sense of personal attachment, maybe it is bunk?

Consider that most of what we think of as history is composed from second hand accounts. There are simply precious few documents extant today that were written at the moment or even close to the time of a "history making" event. There are, for example, no primary, or first hand, accounts of the early childhood of George Washington. It was only after Washington's death in 1799 that Parson Mason Weems published his biography of our first president. In it, Weems recounted the tale of young Washington's truthful admission to having cut down his father's cherry tree. The story caught on and by virtue of being repeated, became "history." Sad to relate, Weems made the story up. But, without primary documents, who could refute him? Who would even want to? Washington was a hero in the eyes of the nation; the peculiar blend of fact and fancy, which we call "history", had secured his position. But what if first hand accounts had been available? Would the cherry tree story still have been accepted and repeated? I think so.

Why? Well, look at Lincoln. A written, published and generally accepted first hand account of Lincoln's inauguration relates that during his inaugural address Lincoln was unable to find space on a small rickety table next to the podium on which to set his hat. This account states that Lincoln's former opponent, Senator Stephen Douglas, reached out and held the new president's hat during the entire address in a gesture of support; "It was a trifling act, but a symbolical one, and not to be forgotten, and it attracted much attention all around me." It was indeed a wonderful story, picked up by all the newspapers of the day, but a story it was. It never happened. Like the cherry tree incident, however, it should have. Maybe much of history is bunk, but the high ideals history sometimes embodies for us are thankfully real and attainable.

Have you ever wondered about families? For some time it seemed as if that was all anyone wondered about. Remember the cant of just few months ago? Does it take a village or does it take a family to raise a child? Despite all the election year rhetoric, despite all the posturing, despite all the noise, the answer is just about as clear as it ever was. And I don't mean just since last November when we re-elected President Clinton. I mean since we elected our first president in 1789.

In Washington's time it was customary for children to leave their families and start on their life's work at an age when our children are still in grade school. Children would be either apprenticed, that is sent away to live in other households in order to learn a trade, or sent to work as servants in the employ of their social betters. In colonial times, the village did indeed raise its children.

Not our idea of what a "family" should do or be? Mine neither. But, in their defense, our notion of "family" did not develop until a bit later, during the decades just before the Civil War. It was during that time that industrialization effectively killed the family oriented cottage industries of the colonial period and "family" came to mean a bread-winning husband, a homemaking wife and children for her to nurture. It was not until the 1920's, however, that the ideal approached reality. Before then, a majority of American children worked an average of 60 hours a week in factories, sweatshops, mines or the households of the well-to-do.

But that was then. Let's get a bit closer to home. Past the 185 million immigrants who came to America between 1890 and 1915 with their children, all working to achieve the American dream. Past the child labor and mandatory education laws of the 1920's. Past the Great Depression, when divorce was down, but desertion was up. Past the roller coaster of World War Two when marriage soared in 1941 and divorces doubled in 1946. All the way to the 1950's when the idea of "family" seemed to have finally materialized. Dad worked, Mom was a homemaker and the kids went to school. Or, at least that was the perception. In fact, the rate of poverty among children in 1957 was higher than it is today, about 30 percent. Teenage pregnancy forty years ago was double what it is today.

So, listen to the pundits. One will tell you that it takes a village to raise a child. Another will tell you it takes a family. But you and I know that it takes both. We've always known that. But I have to wonder; when are we as a country going to be able to elect leaders who also know that?

Have you ever wondered what a blind person dreams about? A person born blind, mind you. Want a clue? Consider your own dreams; I won't bore you with mine. Unless you're a Freudian, you and I probably share the universal conviction that listening to other people's dreams is on a par with listening to a four year old tell you the plot of a movie you haven't seen.

Does the word "disjointed" mean anything to you? Yet, are your own dreams really that connected? Not in the sense of a movie perhaps, but to your life. Of course they are. Somewhere along the line we all learned that dreams are the mind's way of sorting out and coping with all the information and sensations we are confronted with on a daily basis.

Consider how much input you receive each day from each of your senses. Take driving to work, for instance. If you're a commuter, a freeway driver, you are constantly being bombarded with bits of information from your field of vision, your hearing, touch, taste and smell. Now, most of that information is filtered out immediately by your conscious mind as being irrelevant at the moment. You might squirm a bit if your seatbelt is too snug, but you likely don't notice the weave of your pants against your leg. The nerve endings in your leg, however, do. They just don't tell you about it.

All that irrelevant information is stored at an unconscious level, a level where the mind doesn't make such fine distinctions as "real" and "unreal", hence the reason why dreams seem real to us. They are, in essence, made up from pieces of perceived, if not always experienced, information. Want an example? Let's use one we can all relate to; nightmares.

Children routinely experience nightmares in which monsters play major roles, yet how many children have first-hand experiences with such horrible creatures as those that inhabit their dreams? Well, it depends on how you look at it.

To a small child and to the human subconscious, what you see is what you get. A movie monster is every bit as real as the family dog. The experience is there; the monster has been seen and it therefore exists, but is that all dreams are made of? Just visual input? If our dreams are nothing more than rather disjointed movies, a jumble of visual impressions gathered from billions of discrete bits of sensory input over the course of a day, a week, a lifetime, what does a person born blind dream about? At the most important level, the same thing as you and I; feelings. The only difference is that whereas we have pictures, they dream of pure emotions.

My grandmother always said that your eyes are the windows to your soul. It seems to me that it is actually our dreams that allow us to see our souls.

Have you ever wondered what the fastest thing in the universe is? These days, most of us would give top billing to time. Can you think of anything that seems to slip by as fast as the seconds in your day? Me neither, but time isn't the answer. Think back to when you weren't quite so busy. Somewhere back in elementary school you learned that nothing travels faster than light and that light travels 186,000 miles a second. For the science sticklers out there, the actual figure is 186,282.3976 statute miles per second, but 186,000 miles a second is easier to work with, so let's stick with that for the moment.

Now, in one second, make as many complete circles in the air with your index finger as you can. If you're fast, you can probably make one. Try spinning your fingers in eight complete circles in one second. Impossible? Yes, but in that same second, a beam of light could completely circle the earth about eight times. Amazing? You bet. Hard to imagine? You bet. But try this one.

Traveling at 186,000 miles a second a beam of light will travel about 6 million-million miles in a year. In all frankness, I really can't grasp a number like that and I suspect no one else can either, so let's look at it a different way.

Take a peek at your wristwatch, specifically the second hand. Each tick is another second. For your watch to tick 6 million-million times, to equal the number of miles light travels in one year, it would have to have started ticking about 192,000 years ago. Still too large to grasp? I agree. Let's try a more down to earth approach.

Look at your feet. My own size twelves are quite literally a foot long apiece, toe to heel. It would take a beam of light one billionth of a second to travel the length of one of my feet. Scientists call that billionth of a second a nanosecond and while it may not seem like much, it is the reason you actually have more time now than you did when you first learned the speed of light in grade school. In fact, we've gained 20 seconds of time in the past 25 years. How? Well, no matter what it might seem, the Earth isn't spinning any faster these days. In fact it's slowing down a bit. Not enough for you or me to notice, but enough to be measurable in nanoseconds. Those infinitesimal variations need to be taken into account to keep us all in sync, so every once in a while, on either June 30 or December 31, we gain a second. But that's not the whole story.

There's still the matter of how scientists around the world actually calibrate time through the use of cesium atom and hydrogen maser clocks. I'll be happy to explain it to you, if you have a second.

Have you ever wondered how you see? If you are still as dumbfounded by photocopiers as I am, a process that is one heck of a lot easier to understand than how your camera takes a picture, the whole notion of 'seeing' is nothing short of miraculous.

At its simplest, seeing is nothing more mysterious than light rays entering the eye and the message being relayed to the brain via the optic nerve. Amazing, but not miraculous. Not even supernatural, but how about that old feeling that you are being watched? Tell me that you've never had that one before.

Call it power of suggestion if you like, but just writing about it is enough to get me looking over my shoulder, but why? I know I am alone in the house. I know there is no one behind me, but somehow, I feel there is and as soon as I get that feeling, I take a sharp turn off towards the supernatural. The nice thing is that I'm far from alone as I meander down this particular garden path.

Truth be known, a great majority of adults believe that they can actually feel someone looking at them. An equally large number, some 90% of college students, believe that they can actually perceive their own stares. That is to say, they can feel themselves concentrate their gaze on another. If you're thinking what I'm thinking right now, you're wondering how anyone could believe that. Well, the answer is good old experience, our first and best teacher. Simply put, it's happened to us, so we believe it despite what logic tells us. But what exactly has happened to us? We have perceived the world.

As babies we experience the world with our mouths; things come to it. As children we begin to experience the world by touch; we bring

things to us. As adolescents, we begin to order our world by acquisi-
tion, a trait developed more fully in adulthood. Is it any wonder then
that we also tend to believe in a physical connection between our eyes
and the world, albeit a supernatural one? I touch, therefore I perceive.
Why should the eye be any different? Because, you say, Ray Walston—
My Favorite Martian—would influence things by looking at them. We
earthlings have no real physical connection between our eye and what
it sees. True, but many of us think we do.

Let's go back to the creepy feeling of being watched. How many
times have you turned around and found someone really was looking
at you? Not as many times as no one was there, but enough times for
your brain to see a pattern. And each time was an emotional reinforce-
ment of that irrational belief. So, "here's lookin' at you, kid." But,
don't turn around.

Have you ever wondered about the English side of the American Revolution? I realize that sounds rather odd considering that until the war was won both sides were technically English, but ostensibly it was a war between them and us; English versus American, but was it that simple? You know it wasn't.

You know it wasn't because somewhere in your elementary education you were taught that the American colonists didn't have enough to do fighting the English Redcoats, they also had to take on the Hessian troops. What you might not know, however, is just who and what the Hessians were.

First, they were in fact Hessians. That is to say, they were citizens of Hesse, a small state in Germany about the size of Massachusetts. You've probably heard of its capital, Wiesbaden, a resort site since Roman times, but in the 1770's Hesse was known for something else. It was known as the personal dominion of the richest man in Europe, Wilhelm, count of Hesse.

The richest man in all Europe and you never heard of him? Don't feel alone, most people haven't, but I've got another name you will know. Another rich name; Rothschild. It is one of those odd quirks of history that allow us to put these two men, plus King George III and the American colonists all together into one tale. But, let me explain.

Wilhelm was not born the richest man in Europe, he made his money the old fashioned way, he used other people to get it. Specifically he used the able bodied men of his state.

Contrary to popular history, the Hessian troops whom the colonists fought were not mercenaries, that is to say independent paid soldiers, rather their services had been quite literally bought from Wilhelm by King George III for what amounted to a king's ransom. Rich though Wilhelm now was, there was the small matter of converting English pounds, millions of them, into local currency. That is where Rothschild steps in.

Meyer Rothschild, the scion of the international banking family, began his career by supplying rare coins to wealthy clients who collected the coins for much the same reasons as some people nowadays hoard gold; fear. By the time of the American Revolution, Meyer Rothschild was well enough known as a money changer to help convert Wilhelm's pay. It was this boost that led to the eventual House of Rothschild, the backers of Wellington's defeat of Napoleon, the financiers of the Industrial Revolution, the money behind the spread of democracy in the nineteenth century. So this fourth of July, take a moment to think of how much we owe those Hessians.

Have you ever wondered about Circuses? What comes to mind when you sit back, close your eyes and just conjure up your own personal image of a circus? The music? Can't you hear the calliope tooting out the "Entrance of the Gladiators?" No, you read it right. That high-pitched march that you always associate with clowns and acrobats was meant to bring up images of sword wielding gladiators. Incongruous? Not really.

If you look up the word "circus," you'll find that it has its origins in *circul*, meaning circle, and is the equivalent of our word 'circus.' So? Well, close your eyes again and imagine that other figure so vital to anyone's circus memory, the ringmaster. The clue is right there. He is the master of the ring or circle, the circus, and all that transpires within it, but I'd be willing to bet that none of us has ever seen gladiators in our circuses. The ancient Romans, however, did.

In the Circus Maximus, nearly 180,000 people could sit and watch day long spectacles involving men, fierce and exotic animals and even sea battles. So alluring were these spectacles that the poet Juvenal remarked that the poor people of Rome needed only bread and circuses to live. But don't get the idea that the Roman circus was nothing but blood and gore. The most popular sport in the Circus Maximus was the chariot race. The track they ran on was called the hippodrome, Latin for horse course; the name our circus ring still holds today. In fact, can you even imagine a circus without horses? Me neither. Probably because our modern circuses began with them.

In 1710 Philip Astley, an English Calvary sergeant and war hero, traded his commission for a circular track around which he performed daring riding tricks. Eventually other acts were added and thus the modern circus was born in Europe, but America would be the province of a defector from Astley's circus, Bill Ricketts. When Ricketts gave his debut performance in Washington D.C. on April 3, 1793, he had his audience enthralled, including George Washington himself. Washington was so taken by Ricketts that he eventually sold him his beloved horse Jack, the white charger he had ridden during the Revolutionary War. Ricketts did not ride the then 28 year old horse. He displayed him in the first circus sideshow.

Both Washington and Ricketts became media stars of the time, but both fell. Three days after Washington died, Ricketts' amphitheater burned and, in despair, he sailed for England. En route the ship was lost at sea with all hands. A sad ending, but consider what a spectacle it would have made in the Circus Maximus.

Have you ever wondered about television? By now you're probably pretty bored with the summer reruns and the short lived replacements. But take heart. The new season is due to start soon. Question is, how new will it be? Does television hold the same fascination for you it used to?

I have a quite clear memory of sitting on the floor in my house watching "Tom Corbett, Space Cadet" on our old Philco. I can remember being mesmerized by the flickering screen. If you are a baby boomer I'd be willing to bet you also remember The Howdy-Doody club, Crusader Rabbit and his sidekick, Ragland T. Tiger, and Kukla, Fran and Ollie. You might even remember "Thunderbolt the Wonder-colt" if you are old enough. Are you getting the picture? We grew up with the tube. Later generations would have their touchstones also; what child has never seen Sesame Street? But we were there on the cutting edge. Weren't we?

Actually no, we weren't. Were our parents? Well, consider mine. By the time I came around in 1954 they had been married for nearly a decade. About half of that time was lived without benefit of television. Hard to imagine? Again, it depends on when you were born. But a way to visualize it is to look at magazines printed before the television boom of the early fifties. If you look closely at the living rooms pictured in those magazines, you'll notice that each piece of furniture generally faces another piece. If two people sat down, they were of necessity looking at one another. The inevitable consequence of such interior design was conversation and interaction. Compare that with magazine ads from the fifties. Suddenly the entire family is sitting facing a television screen, mesmerized. Is that your cutting edge? The transition from family time to prime time? Sorry, no. We need to go back a bit further. All the way back to the last year of silent movies, 1927.

In that year, twenty year old Philo T. Farnsworth moved to Los Angeles with his young wife and a crazy idea he had about sending pictures through the air electronically. Naturally enough, his neighbors were skeptical about this story and decided he was up to no good. Why did he need all that copper wire? And what was all that apparatus in his back yard for? The obvious answer was that he was making a whiskey still. The cops raided his apartment, but found nothing but an idea that worked well enough to earn Farnsworth a patent later that year for the first electronic television system. Now, seventy years later, Americans watch 1.5 billion hours of television a day, an amount equal to 2,300 human lifetimes each and every day. That's lifetimes spent in inactivity. Scary? You bet. But don't worry, I've got a fancy new replacement for you. Ever hear of something called the Internet?

Have you ever wondered about witches? It's Halloween time again and that means you'll be seeing plenty of them popping up in store displays and in front yards and roofs. Come Halloween night you'll likely even have a number of them come to your door with entreaties for treats. But you know as well as I do that those witches have gotten dressed up in costume for fun. What about the real thing? Are there really such beings as witches? You know the answer; yes and no. Oh, there are people who practice Wicca, an ancient earth religion. They call themselves witches, but deny any evil intent and I believe them, but what of the other sort? The Halloween type? Some people strongly believe in them. Take for instance Jacob Sprenger.

Sprenger was quite famous in his own time for a book he wrote: *The Hammer of Witches*. In our time he is still remembered because that work, first published in Germany in the fifteenth century, has the dubious distinction of having caused more deaths than any other work in history. How? By its definitiveness.

Sprenger, you see, was appointed Inquisitor-General of Germany in 1488. His job, as ordered, was to stamp out witchcraft at any cost. His mission, as he saw it, was to destroy witches, "an evil of nature painted in fair colors." In other words, women. His work detailed not only how to find a witch, but how to secure an unshakable confession from them once found, typically through torture. Those who confessed were executed for the good of their souls. Those who did not confess were executed for the good of the community. But thankfully, there is no dark without light.

Reginald Scot, a propertied Englishman who served in the capacity of Justice of the Peace for his county and so saw a number of witches come up for trial, saw also a disturbing lack of logic in the accusations against them. He questioning nature provided previously missing light. Why, he asked, did "witches" suffer torture and death if they had such power? Why were they almost always desperately poor and alone? Because, he answered, they were easy scapegoats for an ignorant and fearful community. In 1583 Scot wrote his own treatise on witches, *The Discovery of Witchcraft*. This logical study served to somewhat balance the hysteria of Sprenger, but did not annul it. The persecutions would persist and claim an estimated 200,000 people during the 16th and 17th centuries alone, finally coming to a halt in Europe by 1740. But, tell me if you can, do you know so certainly where evil really lies? Perhaps, like beauty and goodness, it is in the eye of the beholder?

Have you ever wondered about circles? Quick now, without thinking—how many circles are in the room you are in? A dozen? You can probably find that many within arms reach. Are you wearing a wristwatch? It likely has a circular face. And consider the buttons on your shirt, they're circular. A hundred? Pretty easy to do that what will all the gears and wheels in the appliances that we rely on. A thousand? Ten thousand? I'm betting you'd get tired of counting before you got them all, but think of how exhausting the job would be if you were outside. How many circles could you find if you were in your back yard? Millions? Sounds way too high, doesn't it? But stand in a single spot and then slowly turn completely around. Within the circle you have just transcribed, how many more circles are there? And just wait until nighttime. Look up at the sky then and consider all the stars that you can see, each of them a dot, a filled in circle if you will, and each of them making their circular paths across the sky as the seasons go by. Millions doesn't sound so high a figure anymore. And if you take a little wider view, you'll see that the stars reside in a greater circular firmament themselves. Circles without end?

It's an old question. The ancient Babylonians were the first ones to consider it. They had watched the moon and the sun make their circles and had fashioned a calendar from their movement, but it was imprecise because the circles were not measurable. The circle made by a draft animal turning a grinding wheel was measurable because it was near, but how could the immense circle of the sun, moon and stars be measured? The answer lay in dividing the circle. The question was how?

Well, the Babylonians did it by using the number six. They counted by sixes and tens, but while there was no easy way to divide a circle into ten pieces, they found a way with six. Let's go back into your yard again. Take a peek at the flowers. Some will have four petals, some five, some six. Pick out one that has six petals. If you look closely you'll notice that the tips of those petals lay on the circumference of an imaginary circle. That is, the flower is circular and the petals divide it into six equal parts. If you draw lines from the tips of those petals to the center of the flower, you will have also divided your circle into six equal parts. If you draw lines from the tips of those petals to the center of the flower, you will have also divided your circle into six equal parts. Keep going and you can divide your circle into as many equal parts as you want. But how many should you want? The Babylonians gave us that answer also; 360, the number of days they thought it took for the sun to travel around the earth, a single year. It seemed obvious therefore that the circle made by the sun as it traveled across the sky each day must ultimately contain 360 of these parts, or degrees. You've probably figured out by now that when they applied this notion of there being 360 degrees in a circle they also garnered the ability to map the earth as well as the heavens. Look at any compass in use today and you'll see the same 360 degrees the Babylonians used 40 centuries ago when they began to expand their empire on the dry, featureless plains of Mesopotamia.

You see, their ability to divide a circle had even more uses than plotting the seasons and knowing when to plant and when to harvest. It allowed them to grow as an empire. They were the innovators of a lighter and thus speedier and more maneuverable war chariot that used spoke wheels—a divided circle—instead of the heavy solid wheels of their soon to be conquered neighbors. It was an idea that gave them supremacy in peace as well as war as they opened trade routes to the Mediterranean civilizations that would learn from them and eclipse them in centuries to come. The Romans, for instance, took their invention of a lightweight war chariot and conquered the world with it and having conquered, civilized. Without the Babylonian idea of a 360 degree circle, the Romans might never have hit on the idea of using the arch, a half circle, for their magnificent buildings, bridges and aqueducts. And without these, there would have been no Rome and thus no Roman Empire. But the story does not stop there. If you'll take a look again at your wristwatch, likely the circle closest to you now, you'll notice that it is also divided into 360 degrees; sixty minutes of sixty seconds each. So, it might have taken 4,000 years, but in this instance at least, it seems that we've come full circle.

Have you ever wondered about Thanksgiving myths? Remember all the Thanksgiving hoopla you had in elementary school? The picture puzzles where you had to find and color the pilgrim's hat buckle or the roast turkey? How about the Thanksgiving Pageant you were in? Remember how you used yards and yards of black butcher paper to make the pilgrim's costumes? Maybe you re-enacted the feast itself, giving thanks to God for the abundance of this new and unknown land. Well, truth be known, the real pilgrims wouldn't have known quite what to make out of your show or most of what we learned about them in school. Time honored and cherished traditions notwithstanding, much of what you know isn't true.

Take that picture puzzle. You might as well quit looking for the hat buckles because there shouldn't be any. Pilgrims didn't wear buckles on their hats. The style only came into vogue towards the end of the seventeenth century. And the turkey? Sorry. Historians note that wild game such as venison was more likely to have been served. And if a domestic fowl did wind up on the table it would have been goose, not turkey. Black clothing? Nope. Black cloth was generally reserved for the wealthy because of its cost. Bright colors were easier to make and maintain; red, purple, maroon, green, blue. Most of the pilgrims would have opted for these colors and made the landing at Plymouth Rock a real pageant. Not that anyone would have noticed. The pilgrims didn't meet their first Indian, Samoset, till three months after they landed. And his arrival wasn't much of a to-do either. He simply walked into the settlement and said, "Welcome, Englishmen." Oh yes, he spoke English, many Indians did, and he knew where they were from because he had met Europeans before. Prior to the pilgrims, the area of New England had been visited and mapped by English, French and Portuguese explorers and fishermen. Matter of fact, the area was quite well known, unlike much of the western part of this new continent. Which brings us closer to our home and an earlier thanksgiving, specifically, April 20, 1598.

It was on that date that Don Juan de Onate, the leader of 400 men, 130 women and children and 7,000 head of livestock reached the Rio Grande River and called for a great feast of thanksgiving. It had taken the assemblage four months to cross the uncharted Chihuahuan desert and now their goal, the area of present day Santa Fe, New Mexico, was within reach.

Historically speaking then, the first Thanksgiving in the new world was not a pilgrim repast after all. But, you know, I still cherish those myths. It seems to me that they bind us together and for that unity alone we should give thanks.

Have you ever wondered about gifts? It's the season again. The time of the year when you brave the malls and try to find the one perfect gift for the special people in your life. Inevitably you get stuck, however, because there is always one special person who really does seem to have it all. But if you think you've got it bad, think of how the three wise men must have felt.

Gold made sense. Whichever of them drew that gift lucked out. But what of the other two? Myrrh? Not a no-brainer like gold, but not too hard to get. It is the resin of any number of thorny shrubs that ancient people used as incense for perfume base. But frankincense? Now we're talking gifts.

Think Microsoft. Consider what it would be like to have a virtual stranglehold on the single most important commodity in your world. Only, we're not talking computer software here, we're talking tree sap. Hard to relate to? Well, consider that in biblical times every altar and every funeral and every temple in the known world needed frankincense as an aromatic offering.

In Egypt and Babylon alone thousands of pounds of the stuff was burnt each year. The Arabs sent a thousand silver talents' weight each year in tribute to Darius, King of Persia; about a million dollars worth in our money. Alexander the Great sent 500 silver talents' weight to his tutor as a gift. Quite a gift! As the ancient historian Pliny, wrote, "It is the luxury of man." But why was it so valuable? Simple. Supply and demand.

The frankincense trees were a jealously guarded treasure of the Hadhramaut people of Arabia. They grew nowhere else and not every

Hadhramaut could harvest them. According to the ancient historian Pliny, it was hereditary and a sacred right. The chosen few would make small incisions in the trees much as New Englanders tap Maple trees today. The sap, once dry, would be sorted by color and size. The best quality, "luminous as moonlight" commanded the highest price, while the poorest quality went for the least, but all of it was destined for the insatiable demands of the ancient world at one hundred times its cost to produce. It would get to its various destinations by trade routes which were as jealously guarded as the trees themselves. For the traders, the penalty for varying from those routes was death, but the secret would not last forever.

About 45 years after the wise men presented their gifts, Hippalus, a Greek, became the first Westerner to discover how to follow the winds and reach Arabia by sea. The Romans followed him. Soon after, the frankincense monopoly was broken and its cultivation begun in new lands, but even then, frankincense would long remain the true gift of kings.

Have you ever wondered about language? You have you know, but you probably don't remember. Think back to a time when you were very, very small. For myself, I vividly remember finding a most wondrous toy wooden train at a gift shop down at the Pike in Long Beach, right about where the Aquarium of the Pacific now stands. I picked it up and started looking for my mother so that she could buy it for me, but she found me first. She took the train from me, put it back and explained to an amused sales lady that I was too small to understand that it had to be paid for, that I thought it was free. I, on the other hand, was mortified because I knew it had to be paid for, but I didn't know how to say so because I couldn't talk yet. I've long wondered if perhaps that event led me to choose writing and teaching for my life's work.

Language and learning first went hand in hand for me that day, but you and I can both cite times they didn't. We've all had droning teachers and mind-numbing reading assignments. The prime example is high school English and Shakespeare. Can't you still hear the groans when your teacher brought out her ragged old paperback copies of <u>Romeo and Juliet</u> and passed them around? You sighed as your teacher told you what a great play it was and how much you could get out of it. Right? All you had to do was try to understand it. Remember the first question? "Did people really talk like that?" Do you remember her answer? "Yes and no."

Any author wants first and foremost to be understood, so he uses language his reader will understand. But beyond that, a writer often uses his words to instruct. To that end, his writing must also be better than ordinary language; it must be memorable. But, you say, the language you were forced to read in school wasn't memorable. Otherwise, you'd remember the words, not the assignment. Well, try this on for size; "the spirit is willing, but the flesh is weak." Familiar? How about when you felt you were living off the "fat of the land." And which one of us hasn't asked, "am I my brother's keeper?"

You see, you do remember the language, some of it anyway, because it spoke to you. But, no, it isn't William Shakespeare. It came before him. It's William Tyndale, the English priest who in 1526 became the first to translate the Bible from Greek into common English and was executed for his seditious act. Even then words had power and those who knew the words wanted to keep the power. Such a monopoly couldn't last, but the revealed words would. Although he is not as revered as Shakespeare or as much remembered, Tyndale's writing had a more profound influence on his time than Shakespeare's on his and his language still reverberates today. Although few attain it, that is the legacy every writer desires; to not be the solitary "voice crying in the wilderness", but rather to be as vital as a raindrop and as far-reaching as a ripple in a puddle.

0-595-22601-9

www.ingramcontent.com/pod-product-compliance
Lightning Source LLC
Chambersburg PA
CBHW030753180526
45163CB00003B/1004